© 2024—Mimesis International
www.mimesisinternational.com
e-mail: info@mimesisinternational.com

© MIM Edizioni Srl
Piazza Don Enrico Mapelli 75
20099 Sesto San Giovanni (Milan, Italy)
VAT: (IT)02419370305

From an idea by Roberto Petronio

Author: David W. Brown
Interviews edited by: Andrea Biancardi, Andrea Conconi, Clelia Iacomino, Mattia Pianorsi and
Aristea Saputo (SEE Lab, SDA Bocconi School of Management)
Interviews coordination: Simonetta Di Pippo (SEE Lab, SDA Bocconi School of Management)
Project coordination: Ivana Giannone (Telespazio) and Aristea Saputo (SEE Lab, SDA Bocconi School
of Management)
Supervision and image processing, relations with the publisher: Sara Udinesi (Telespazio/e-GEOS)
Editorial consulting: Paolo Mazzetti (Telespazio)

Publishing coordination: Grimaldi & Zevi, Milano

Paperback ISBN: 9788869774676
e-book ISBN: 9788869774683

Layout and Cover: Valeria Zevi
Editor: Caterina Grimaldi

Cover Image: The Amazon River, graphic processing by e-GEOS using images from COSMO-SkyMed © ASI

SPACE
IN OUR HANDS

David W. Brown

SPACE
IN OUR HANDS

MIMESIS
INTERNATIONAL

Cont

ents

Preface

Two sides of the same coin, inextricably linked. This is the first image that comes to my mind when the concept of sustainability is brought up. The idea, in its most classic sense, is generally expressed as actions to protect our planet. However, the technologies available to us, our ability to colonize outer space and our inclination to move towards other celestial bodies require us to rethink and expand our horizons, to the point of encompassing space within the concept of sustainability. Space is no longer exclusively of interest to scientists, it has the power to affect our lives on Earth. We observe the Earth using sensors with different technologies, which extract mutually complementary information, and we process these data with algorithms that can produce more refined and accurate information. This all translates into a type of knowledge that the human species has never known before. It is used to serve people across the most diverse range of sectors, including emergency management, precision agriculture, the protection of the environment, the territory and the seas, right up to fighting to contain climate change. It is a lot, but still it is not enough. We need to move (as we are actually already doing) from the ability to monitor to the capacity to predict and, therefore, to act before the damage is visible.

So if it is true—and true it is—that our ability to change direction and fight the effects of climate change depends mostly on satellites, how can we not think about looking after the environment in which satellites move and operate? How can we imagine we can continue polluting and overcrowding outer space, jeopardising the progress we have achieved with so much effort and the ability of future generations to access space?

It is clear: we can't. This is precisely the starting point of this book, which stems from the conviction that space has given and is still giving so much to humanity, also and especially in terms of tools for sustainability, and it is now time for us to give something in return. In other words, we should try to avoid, with outer space now crowded with tons of junk, making the same mistakes that we are trying to remedy on Earth.

Just over a year has passed since we in Telespazio became convinced that the time had come to present these considerations to a broader audience, to far more people than the ones who so often have heard me speak about these concepts. To do it in the best possible way, we needed three things: a high-level partner that believed in this project as much as we did, voices that could speak with authority (and explain a complex world in simple terms) and an author who could put these voices together and make them understandable to the lay public as well. We were lucky to find all three.

Our collaboration with the SEE Lab of the Bocconi University has for a long time been one of our strongest partnerships. I would like to thank this prestigious institute for taking part in this new project and to have provided very valuable ideas and contributions.

A completely new but equally successful partnership is our work with David W. Brown. For a far-reaching book, distributed in Europe and the United States, we needed an author who could captivate readers and explain. With his considerable experience contributing to the most famous American papers and a successful best-selling book, David was immediately the natural choice.

Paolo Nespoli, Simonetta Di Pippo, John C. Mankins, Paolo Gaudenzi, Waltraut Hoheneder, Barbara Imhof, René Waclavicek, Angel Abbud-Madrid, Kevin O'Connell, Moriba Jah, Cynda Collins Arsenault and Victoria Samson generously accepted to add their knowledge and professional skills to the pages of this book.

What we wanted to collect in this book was a new vision of space that will apply in the coming years and decades, when working and living on celestial bodies other than Earth will no longer just be science fiction. It is a challenging prospect, which reminds us of our responsibilities and serves as a reminder that, like Earth, space is in our hands.

Luigi Pasquali,

Preface

Consistently with the scientific community's repeated claims on last years' rising temperatures, 2022 has been scattered with extreme climate events all over the world. Just to mention a few, Europe experienced severe heatwaves, increasing wildfires' emissions, and spreading droughts due to reduced rainfalls. India and Pakistan faced intense heatwaves followed by floods hitting millions of people and hundreds of thousands of livestock. In China, the Yangtze River areas were affected by intense droughts before some floods inundated them and other regions, and in the United States heatwaves, droughts and floods occurred too.

At the same time, the concentration of carbon dioxide and methane in the atmosphere have kept increasing with respect to previous periods, and spreading Covid-19 and geopolitical tensions threatened, among others, global energy and food supplies, exacerbating existing economic and social inequalities within and across countries.

While on Earth the need of redesigning production and consumption models to face global challenges has never been clearer, the sustainability of our activities outside the atmosphere is rising concerns as well. In particular, orbits are growingly populated by space debris, abandoned unutilized materials that, wandering out of control, might hit operational spacecraft, damaging them or definitely shutting them down, with major consequences for the services that they provide. While the space sector has long been left aside by sustainability regulations due to its relatively limited production rate, current growth scenarios and technology developments' potentialities are inducing some industry leaders to question their own role in preserving Earth's ecosystems and the space resources upon which they rely.

Other than scrutinizing potential solutions to its own footprint, the space industry is actually already contributing to transform terrestrial activities. Earth observation, satellite telecommunication and navigation, space solar power, research initiatives conducted in orbiting space stations and during exploration programs, are all proving to be crucial for addressing the United Nations Sustainable Development Goals (SDGs). Today, by collecting data from Earth's orbits, providing connectivity to remote locations, and testing new technologies in the space environment, space is offering a key contribution to humanity's most urgent needs, from ending poverty and inequalities to sustaining better health and education conditions, from triggering economic growth to achieving environmentally and socially sustainable industrial processes.

In this context, being aware of the space sector's potentialities might be key to everyone to advocate new solutions for the humanity's sustainable transition. This book not only provides a comprehensive and engaging overview of the most salient interconnections between space and sustainability, but also does so through the words of internationally recognized experts on the topic, who are dedicating their lives to better understand how space can support humankind face its current and future challenges within and beyond the Earth's limits.

I trust that a shared understanding of space-related opportunities might enable new solutions and let them thrive, sparking awareness and creativity, so I am happy of having jointly developed this project with Telespazio. I have always been striving for a more sustainable planet and I hope that this will be true also for all future readers who will let this book inspire them.

Simonetta Di Pippo,
SEE Lab Director, SDA Bocconi Professor of Practice

A New
for
and Sus

Philosophy
Space
tainability

As a matter of biology, the human race had all it needed to build a spaceship about one hundred thousand years ago, during the Middle Paleolithic. That was when we evolved at last the big brains to drive our upright bodies.[1] Still, it took ninety thousand years before we would work out both animal husbandry and agriculture, let alone spaceflight. At the time, there existed no known way to preserve hard-won expertise. What wasn't spoken aloud was lost forever. Knowledge was a war of attrition.

With farmland and beasts of burden, however, came the blessed by-product of stability, and within five *millennia*, the first written records emerged. It was, perhaps, the fastest technological breakthrough to take hold among humans, and the most important. Soon, the wars we once waged with the natural world were directed within, and empires emerged and were extinguished by way of weapons of bronze and then iron, until somehow, in perhaps the most fortuitous stroke of luck, our beautiful race of art and ferocity forged, at a small school in Athens, the concept of philosophy.

Given the merciless world with which our frail, spear-clutching species had to contend, one hundred thousand years from the brain to Plato's *The Republic* represents real rapidity. All proceeded from there, and Socrates, Plato, and Aristotle gave us the tools to build the future.

For all our achievements, however, the 500 days beginning on 20 June 1944 mark the single most consequential span of time in human history.[2] From then on, nothing would ever be the same again, and if the convergence of art, science, and engineering were not wielded with the utmost care and respect, there might be no future for our species, or, indeed, any creature of the Earth.

In those 500 days, humanity sent into space the first man-made object, the German rocket MW18014; we applied the work of physicists Fermi, Chadwick, and Rutherford to practical means, and detonated the first atomic bomb; and we switched on the first electronic computer prototype.[3] With that trifecta, we witnessed at last the terrible power of the smallest size—the atom, a philosophical concept that comes to us from ancient Greece. We recognized that we possessed the tools to travel to the cosmos—to infinity itself. And had now at our disposal a means to calculate these things astoundingly large and inconceivably small. Humanity's full potential took one hundred thousand years to reach, and 500 days to realize.

Today, things quite clearly are worrisome, and at every level, the common person feels unmoored and abandoned by political figures and business interests, unsettled by institutions once thought benevolent, and disappointed in an international order that once promised a Pax Humana just around the corner. Any progress that might once have registered on the United Nations 2030 Agenda for Sustainable Development was slowed down by the COVID-19 pandemic. By all accounts, things seem likely only to deteriorate further from here, and if history is indeed any guide, days dark as the 1930s seem set to engulf us. But if that is so, the universe that was opened to us in the late 1940s, when the war ended and the greatest atrocities were halted, and its monsters were brought to account, might yet be our salvation.

*

There is no human endeavor better encapsulating the philosophy of sustainability than the exploration of space. The human and robotic expansion ever deeper to the stars is foundational for a better now and enables a better next.

Anything less than a total commitment to these goals, by definition, sacrifices the present. In the extreme long term, multiplanetary settlement may preserve human civilization. In near term, space-based platforms improve human interaction through orbital communication and navigation, and monitor the planetary health of the Earth's climate, sea life, and cryosphere. This is to say nothing of the vital nourishment of the human spirit through peaceful exploration of the unknown.

As a matter of terminology, the Brundtland Commission of the United Nations defined "sustainability" in the 1987 paper *Report of the World Commission on Environment and Development: Our*

Common Future, describing it as the ability of "meeting the needs of the present without compromising the ability of future generations to meet their own needs".[4]

The definition of the planet Earth is well known, but our collective understanding of space is generally mistaken. Space is not something *out there* from which we are wholly removed *down here*. The objects in our solar system—the "things in space"—are not even independent of the blue marble we call home. The Earth is literally inside the Sun's atmosphere. Mars rocks litter the Earth, as does debris from the asteroid belt and chemical signatures of comet strikes.[5] (Markers on Earth of our celestial interconnectedness might include such small things as... the oceans—and possibly life itself!) And so great was the asteroid impact 65 million years ago that it likely kicked massive debris into deep space, which means intrepid astronauts one day might dig up dinosaur bones on the Moon.[6] As Sir Fred Hoyle quipped, "Space isn't remote at all. It's only an hour's drive away if your car could go straight upwards".[7]

What, then, is space? It is where every human being now lives. The Earth is in space, circling the Sun 30 kilometers per second, in a solar system hurling across the galaxy at 220 kilometers per second, the galaxy itself crossing the cosmos at 580 kilometers per second. The influence of space affects every creature and system of the Earth because we are inextricably linked.

The division of "Earth sustainability" and "space sustainability" is therefore specious. They are one and the same, necessitating conception of a new, integrative, holistic model of "sustainability". Such is the prerequisite for true progress for the creatures on Earth, and our slow but advancing incursions ever farther from the planet's surface.

Petty political problems threaten the increased separation of humans from planet Earth. Satellite-destroying missiles yield debris that can render low-Earth orbit utterly inhospitable to future spacecraft at that altitude. At eight kilometers per second, even a small bolt can act as an armor piercing shell on another spacecraft it encounters. Satellite mega-constellations bring to developing countries and the distant corners of the Earth broadband Internet access and the bulk of the human race's collective knowledge. If they are inadequately hardened to the caprices of the Sun and its flailing corona, however, they, likewise, can crash and cause cascade effects that render low-Earth orbit inaccessible.

The Sustainable Development Goals (SDGs)[8] of the United Nations cannot be achieved without the incorporation and integration of space sustainability, and space sustainability cannot likely be achieved without such organizing institutions as the United Nations.

Exploration is a manifestation of our innate sense of curiosity. Never in history has our curiosity been sated, because curiosity itself is driven, at least in part, by the desire to minimize the pain of existence. We search for new fields to forage, new animals to subdue, and new land to cultivate, each alleviating hunger. We search for rivers and wells and springs and streams to quench our thirst. In mountain, forest, and creek, we seek ore to refine, wood to build, metal to fashion. Equal to our material needs, we seek out the transcendent. Did the mountaineer atop Pikes Peak, upon spotting massive herds of migrating bison, see food, or see the splendor of the natural world? Has anyone ever seen a waterfall and thought merely about drinking it? No Apollo astronaut pressed his boots into the lunar surface, gazed on the alabaster horizon, and felt grateful for the untapped veins of palladium beneath his feet.

Humans have in the course of exploration demonstrated shocking short-sightedness. Those bison, for example, were hunted nearly to extinction. But humanity over time has often displayed a unique ability to orient its goals such that our quest to minimize pain today does not compromise ourselves a year or a generation from now. The bison did not go extinct, and today are not categorized by the United States National Parks Service as a threatened or endangered species.[9]

Exploration is painful, but is also an endless drive to ameliorate greater collective pain, and as a species we have pushed relentlessly out of one wilderness and into the next. The spiritual satisfaction derived from our endeavors is not in their completion, but in their pursuit. This is consistent with the assertion of Albert Camus: "This universe henceforth without a master seems to him neither sterile nor futile. Each atom of that stone, each mineral flake of that night filled mountain, in itself forms a world. The struggle itself toward the heights is enough to fill a man's heart. One must imagine Sisyphus happy".[10]

The exploration of the infinite is as urgent in both directions—outward and inward. The telescope was invented almost immediately after the microscope. We seek to explore the constituent parts that make up the human creature, first cells and then nucleotides and then atoms. We seek to explore the constituent parts that make up our solar system, and by implication, that make up our world. The questions that drive science might include: What are we? Where do we come from? How do we fit into a cosmic whole?

To find those answers, we build, among other things, electron microscopes, mountaintop telescopes, and perhaps the grandest of them all, spacecraft.

Humans are uniquely capable of balancing the needs of today with the needs of the future, and at leveraging our curiosity toward materially productive and ultimately conservational ends.

After the discovery of the ozone hole over Antarctica, NASA mobilized the Total Ozone Mapping Spectrometer and Solar Backscatter Ultraviolet instruments, carried by multiple spacecraft, to determine the scale of the problem.[11] Within two years of the ozone hole's discovery, 197 countries signed the *Montreal Protocols*—the only United Nations treaty in history to achieve universal ratification.[12] It was perhaps the first and best example of how holistic sustainability of the terrestrial and celestial could be not only accepted by countries as apathetic to one another as North Korea and the United States, but bring them together for geopolitical change.

As Earth scientists, planetary scientists, mathematicians, and astronomers discover ever more issues of sustainability, including those described in these pages, the *Montreal Protocols* and the terrible discovery that led to their drafting is the model for the future.

*

The way we would save ourselves from ourselves began, like so many things, with Galileo Galilei. At the close of the sixteenth century, Galileo was the chair of mathematics at the university of Padua in the Venetian Republic. He took the post after having distinguished himself at the University of Pisa, where in the area of falling bodies, he disproved Aristotle of all people, which is not a bad start for any professorial career. But in Padua, he would really make his mark. He determined through experiment that the density of a fluid is related to temperature, and years later, would expound a bit further on this in his final book, *Dialogues Concerning Two New Sciences*.[13] This principle formed the foundation of what would later be called the thermometer.[14]

This little slice of science trivia would later affect the story of medicine, itself having been one long and oftentimes ugly slog through the unknown. Though we live now in the best time in human history from a standpoint of medical care and practice, one hundred years hence, people will undoubtedly look back on us as though we were barbarians, and our doctors a minute removed from the bloodletters of old. Such is the nature of medicine, and so much the better for the prospects of humankind.

Galileo's indirect influence on the practice of medicine occurred one hundred years after his early research into proto-thermometers, when the first physician used a precision thermometer in clinical studies.[15] This kicked into motion an eventual quantum leap in medical care: the concept of "vital signs".

Today, physicians and nurses keep close watch on four key measurements of a patient's health: temperature, pulse, respiratory rate, and blood pressure.[16] Ongoing checks of vital signs over the course of a hospital stay, and across repeat outpatient visits, improves health outcomes and offer early warnings of trouble ahead.[17] Though the concept took time to adapt, and the specific measurements remain in flux, today it is axiomatic that to sustain a patient is to know his or her vital signs.

Galileo had no way of knowing any of this (and indeed, his ther-
moscope would have been utterly useless to the practice of medicine), but such was the breadth and genius of the Florentine's research that even his dabblings can somehow bring together fields
of science and human endeavor in meaningful, almost transcendent ways.

Albert Einstein called Galileo "the father of modern science", and
indeed, just as the "stars of the Medici" Galileo discovered did in
fact orbit Jupiter, and prove the hypothesis of Copernicus, so too
does a notion of space sustainability orbit Galileo.[18] The vital signs
of medical patients have a direct analog around Earth.

We live in an age and perhaps epoch of global turmoil, not simply
in terms of civil strife, but also as a matter of upheaval and insecu-
rity in the planet itself. The Earth is changing and we do not know
why. To mitigate the uncertainty, we need to do what doctors do
for their patients. We need to take the planet's vital signs—and
there is only one way to get them.

Earth is about 4.5 billion years old. Only in the last sixty-five years
have denizens of our world been able to reach beyond it. We have
done so in ways productive and not, but one thing we know for
sure is how to take measurements from Earth's orbit of the planet
below, and use them to diagnose problems both ongoing and new.
With that information, scientists can advise policy makers on
how best to cure the patient. Today, about six thousand satellites
circle the Earth. Most are space junk—old dead spacecraft whose
orbits have not yet decayed. About 162 of them, however, can be
classified broadly as "climate satellites".[19]

Determining the heath of planet Earth, goes far beyond an altitude of one hundred miles however. The exploration of deep space
will likewise yield data vital to the sustainability of Earth. Beyond
the measurement of the Earth's vital signs, it is imperative to study
those of other worlds. This scientific endeavor, called comparative
planetology, involves the employment of telescopes and spacecraft to take key geologic, thermal, atmospheric, meteorologic, and
chemical measurements, among others, to understand how planets work. Such knowledge feeds directly back into the understanding of the Earth, enabling a better comprehension not only of the
mechanics of planets, but also allowing scientists to constrain the
effects of human beings on our world. Moreover, such knowledge

will benefit future Mars-born humans, who will be able to practice an environmental stewardship more responsible than perhaps that which is practiced on Earth. On millennial timescales, should terraforming become a reality, the field of comparative planetology will likewise allow that enterprise a greater chance of success (or at the very least, a lower likelihood of catastrophe).

The key to all this is a robust and ongoing investment in space exploration both robotic and human, with buy-in from government, academia, and industry.

We have seen the effects of a massive, rapid infusion of capital once before: During the Space Race between the bitterly rivalrous United States and Soviet Union, in fewer than twelve years, humans went from launching a small artificial, Earth orbiting satellite to landing human beings on the Moon. This, one might fairly assert, was a historical anomaly, enabled by a unique confluence of a reorganized global order following a devastating war, the emergence of technologies of unbridled potential, and paradoxical national psychologies willing to execute missions impossibly dangerous. In other words, the space programs of the two superpowers were fig leaves for the development of weapons of mass destruction unlike any the world had ever before considered, let alone built. But we only launched the astronauts and cosmonauts. The fusion bombs never left their silos.

The support structures of the Space Race were untenable in the long term. When three astronauts died during a rehearsal for Apollo I, the sense of the dominance of engineers over nature sustained a body blow in the public conscience.[20] (Americans and lawmakers were willing to marvel at the heroism of men who would strap themselves to missiles to best their Soviet counterparts, but were less willing to watch men commit suicide on the public dime, or worse, allow carelessness to take the lives of heroes.) The financial burden of the Space Race likewise exceeded the appetite of the body politic, particularly as the stakes of the American-Russian rivalry changed. The Cuban Missile Crisis cured many of their affection for nuclear weapons. The space programs of both countries were *de facto* development programs for intercontinental ballistic missiles, and by Apollo 17, their utility in that regard had diminished. The entire enterprise, in short, was a fast, spectacular sprint as opposed to a lengthy, unremitting marathon.

As matters of technology development, private investment, and, with the signing of the Artemis Accords,[21] renewed international involvement in settling deep space, one may reasonably argue that today, rather than the 1950s and 1960s, is the Golden Age of space exploration.

On an extremely long timeline, one may assert that the period beginning in 1957 through today constitutes a single, uninterrupted Golden Age, but the move from government to industry in spaceflight hardware responsibility is a clear paradigm shift. The durability of present spaceflight advances, however, is not assured.

No humans, for example, have yet died on a private rocket launch, though such an outcome is inevitable over time. After the Apollo I catastrophe—not even an actual launch—no humans were launched for twenty months afterward, then the longest such delay for an active human spaceflight program. After the Challenger disaster, NASA launched no one for two years. Again, after the loss of Columbia on reentry, two years elapsed with NASA unable or unwilling to launch humans.[22] Eventually, NASA ended the Space Shuttle program unceremoniously, without even bothering to have a successor launch system in place.

Governments can endure such long pauses in launch cadence. It remains unclear whether the private sector would likewise be able to maintain operations and development during subsequent investigations and a possible stand-down. Moreover, such a disaster might invite stricter government oversight and regulation, which might again slow or end the pace of private space sector development.

Though entrepreneurs have tried with various levels of success to manifest new use-cases for space, what might have seemed like obvious potential businesses fizzled and failed. Planetary

Resources, Inc., for example, backed by billionaires and which sought to mine raw materials from asteroids, collapsed while waiting for the private launch industry to mature. The same fate awaited Deep Space Industries, another asteroid mining outfit, and Shackleton Energy Company, which sought to mine the Moon. Resource extraction has obvious profit potential. The same gold, silver, palladium, diamonds, and uranium found on Earth exist on asteroids and other celestial bodies. Helium-3, an isotope that would greatly enable fusion power, is abundant on the Moon and virtually impossible to access on Earth, where its greatest abundance is in the planet's mantle.

This is all to say that no one knows, truly, the durability of our Golden Age, and whether, if struck by a devastating setback, the age would be brought to an immediate closure.

What one can say, however, is that at present, kinetic operations by our new public-private partnership have reached a full gallop. In an exceptional month, NASA's Space Shuttle program could launch two shuttles. SpaceX can launch three rockets in less than two days.[23] If this continues to be done in a responsible manner—and rapid reuse of rocketry is unambiguously a worthy endeavor, both from environmental and global economic standpoints—there is virtually no limit to what can be moved from Earth to space. It is ten times cheaper to launch on a SpaceX rocket than NASA previously paid for expendable launches—and SpaceX accommodates clients around the world.[24] Many nations once excluded from our spacefaring future now have a way of reaching the stars.

Bringing the discussion full circle, space sustainability is pointless if humans kneecap the space program in its present iteration in the name of sustainability. Because this new paradigm is in its nascency, now more than ever is the time to introduce sustainability to the conversation in a way that is organic, and into business cases as they are being written. An industry that internalizes sustainability on Earth, in orbit, and on other worlds, is an industry that can thrive responsibly.

Ironically, though government intervention is cited by some as the solution to perceived failings of the private space sector, the most monstrous demonstrations of destructiveness in space have been perpetuated by government entities themselves. The International Space Station (ISS) must regularly change its altitude to account for space debris. Similarly, hundreds of rocket stages have poisoned parts of the ocean and disrupted local ecologies for decades.

*

Space sustainability can be divided into four major areas, to be explored in detail in this book.

First, spacecraft in orbit around the Earth devoted to monitoring the health of the planet are the cornerstone of sustainability. They are intrinsic to the notion that space is a net benefit to humankind.

In the year 2000, atmospheric chemist Paul J. Crutzen and biologist Eugene F. Stoermer published a letter in the International Geosphere-Biosphere Programme's *Global Change Newsletter* asserting that the indelible mark of humankind on the planet Earth has thus merited a new geologic era: the Anthropocene.[25] From urbanization to the expenditure in only a few generations of the "fossil fuels that were generated over several hundred million years"—to say nothing of the consequent release of carbon compounds into the atmosphere—the Earth has changed in ways it could not have if humans had never climbed down from the trees. The term has not yet attained any official status, and the precise demarcation points (e.g. whether its beginning was the Industrial Revolution or the Atomic Age) remain debated.[26]

This new epoch necessarily requires study. For broad swaths of Earth science, this means orbital assets looking down on our planet from space, measuring changes, and returning data for analysis. "Sustainability" is a slippery word because so much of it hangs on intention (see the unanticipated effects of everything from diverting waterways to ecological peril, to the gross destruction of the natural ecosystem of Yellowstone National Park during the twentieth century, in the name of scientific management). *But unambiguously, the sustainability of the Earth requires hard data. It requires space.*

Next, spacecraft at Lagrange points and in orbit around the Earth, all devoted to communication, navigation, agriculture, and space weather, are likewise examples of sustainability writ large, because the international comity that comes with greater interaction is a prerequisite for our species to endure.

But everything from the power grids of the Earth to our navigation—even homing pigeon racing—are at the mercy of the Sun's caprices. Coronal mass ejections are monitored in the United States by the Space Weather Prediction Center, whose reports allow mitigation of measured magnetic variation inbound from our star. Here, space protects the hard-won gains of human progress. In orbit, meanwhile, the connection of societies has allowed the exposure of atrocities from inside repressive regimes and war crimes committed on the battlefield. Families separated by oceans may now remain connected to home, and gamers, to name one subgroup, form friendships as real and true across oceans and continents as might be had with an elementary school classmate. Foreign nations were once understood by foreign populaces only by the faces of antithetical regime leadership. Americans once knew Iran, for example, in a way disconnected from the Iranian people, and accordingly, held frightful views of its entire populace. Arguably because of connections facilitated by communication satellites, views have improved. Iranian people, Americans can now see, are people, and public opinion has shifted decisively toward amity.[27] Such orbital assets as global positioning systems moreover allow a confident freedom of movement once out of reach, and the same systems may be applied to technologies beyond navigating roadways. Major agricultural enterprises are driven wholly by GPS-guided tractors, fertilizers, and harvesters. *Space systems are feeding the world.*

Third, there is space sustainability concerning Earth's orbit. Here, most pressingly, the issue of space debris must be met dead-on.

The material poisoning of low-Earth orbit could render vital orbital altitudes utterly inhospitable to future spacecraft, and cripple technological advances that might be applied to the aforementioned climate and communication spacecraft that make life on Earth worth living, with a future of increased opportunity and as-yet unimaginable progress and growth.

Oftentimes, actions antithetical to the environment, whether terrestrial or celestial, by governments and industry are caused by shortsightedness, ignorance, or benign neglect. All human endeavors have always been and shall forevermore be plagued by such second-order and tertiary maladies. Space debris at its most malevolent, however, is an intentional assault on a natural region of the Earth. Every person involved in the design, assembly, testing, and launch of an anti-satellite weapon is fully aware of its consequences. The United States, former Soviet Union, and China have each tested anti-satellite weapons, releasing, cumulatively, shards of debris in the thousands, in massive clouds, many of which remain in orbit for years. The Russian anti-satellite test in 2021 released a debris field of fifteen hundred parts trackable from Earth.[28] More malevolent was this launch, as that nation had already established its ability to destroy a satellite from the ground. The danger of space debris unchecked cannot be overstated. Already, rocket launches must be timed to avoid debris fields, and the Space Station must be raised and lowered to protect it from debris (a practice not always successful—the Space Station has repeatedly taken damage, including complete punctures, from junk in orbit).[29] Of all the space sustainability issues that must be addressed, this is the one that could end, or certainly constrain, any attempts at placing spacecraft in orbit or launching crewed missions to other bodies. *All proceeds from addressing the debris issue head-on.*

Lastly, the principles of space sustainability must be extended to humanity's expansion to celestial bodies beyond Earth.

This is so for at least two reasons. First, one common criticism of space exploration by citizens not invested emotionally or intellectually in the enterprise is that we should not go forth and harm another planet after having harmed this one. It is therefore essential that sustainability drive multi-planetary exploration—to disarm a potent weapon that could bring such exploration to a needless and self-defeating halt. Secondly, the exploration and ultimate settlement of other planetary bodies such as the Moon or Mars necessitate a kind of hyper reuse of everything brought from Earth. In the near term, or certainly our lifetimes, a Mars outpost may survive only by external means. Virtually every molecule brought there must be recycled in some way to better position colonists for self-sufficiency, or failing that, greater self-sufficiency. An utter commitment on the Moon and Mars to recycling materials and biomasses, as well as the conservation of water and protection of air, is precisely the sort of thing one who is environmentally minded might wish to bring to Earth, where the manufacture of virtually every resource in some way likely harms the ecosystem, from the initial extraction of minerals to the chemical plants that turn carbon and hydrogen into plastic. It is a use case where development work done for the settlement of Mars has direct applications on Earth, and moreover, establishes the sort of colonial footprint off world that keeps in best ecological practices on this one.

Space sustainability then is at once a set of philosophical guidelines, a collection of best practices, and above all, a mindset.

It can be motivated by a love of the natural world and the wonder of our cosmos, or by selfish means—preserving a newly accessible economic stream with unbridled potential with only a bit more development and a little more time. There is no human alive who does not have something to gain from space sustainability.

This book shall examine space sustainability across multiple domains, including the ways that it directly benefits the health of the Earth and enables new pathways of progress for quality of life here, as well as the way it should be integrated into humanity's tentative first steps toward establishing itself as an interplanetary species. This text features interviews with researchers and entrepreneurs involved in lunar exploration and the ways it benefits Earth; the closed recycling systems of deep space habitats and potential applications on the Earth's surface; and the implications of space mining on the Moon, Mars, and the asteroid belt. Lastly, the book provides commentary on contextual frameworks for how we as a species move forward vis-à-vis space, and how in making those movements, we secure the ability of future generations to do the same.

Space
for

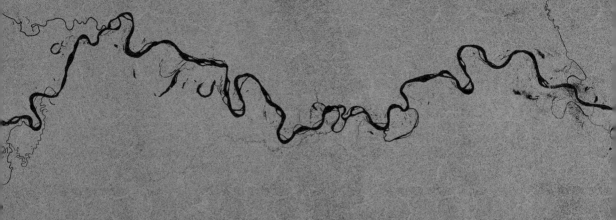

Research
Earth

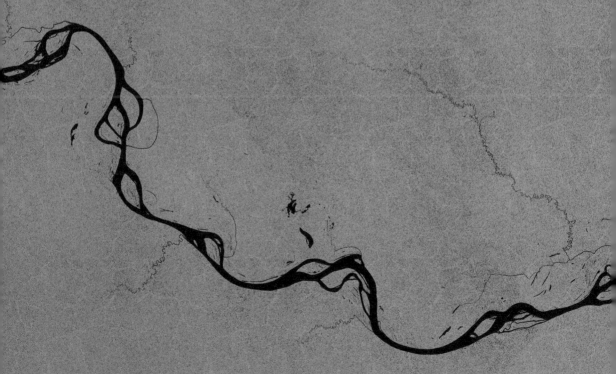

Amazon River 1 (detail). COSMO-SkyMed Image © ASI 2023. Distributed by ASI and processed by e-GEOS

As a human being, life in the Anthropocene is just about as good as it gets. What the world does not provide, we create. We have done this for hundreds of thousands of years, axing trees, carving waterways, and drilling deeply. Only in the last few decades have we had the technology to study this planet from orbit and take from it a broad view of what is happening where, and tease from those data the reasons. And in the same amount of time, we have likewise learned how to study up close other planets close by. From a planetary science perspective, the mysteries of the Earth should be falling away like the rock we've hewn from mountaintops. But during this critical period of a changing climate, there is a frustrating variable that prevents scientists from fully understanding what is happening, and why. That variable is humanity.

When puzzling out the behaviors of the Earth's weather, climate, ocean current, and glacial systems, among others, it is nearly impossible to know which variations are caused by human factors, and which are simply chaotic outcomes from a system of astounding complexity.

Mathematicians can model the climate, for example, and produce reasonable cases of how humans affect the Earth, and how ongoing activities might harm or help it a hundred years from now, but it is undeniable that modeling the atmosphere alone, based on historical data, is an absurdly challenging task and can only with a low level of confidence produce results very far into the future. Likewise, modeling the ocean is an astoundingly difficult if not impossible task, even working with centuries of data from sailors, whalers, lighthouses, and navies. Understanding the cryosphere is no easier task. To combine the three—air, water, and ice—each nearly impossible to model, into a super model with precise outcomes, is a heroic endeavor.

And perhaps impossible. The moral calculus of coercing the behaviors of developing nations based on such a model is thorny at best. Can one reasonably consign poor nations to starvation just a bit longer because dirty power plants might contribute to a fraction of a degree of global warming in a model that is less than perfect? It is perfectly easy to do when you live in the developed world and removed from the human cost of such calculus.

The goal should not be to disregard such models or build as many coal-powered plants as possible. Rather, the goals should be to better constrain the model and feed it more data. In a sense, the climate puzzle is a grand experiment we are performing on ourselves, and, paradoxically, suffers both from the observer effect, and also from an inability to actually observe anything at all.

This is where the exploration of space can help us understand planet Earth. Our adjacent terrestrial planets, Venus and Mars, are very different from Earth—but not that different.

Venus, for example, is about the size of Earth. It's made of the same stuff. It has the same density and mass. Gravity there is about the same, and it has an atmosphere and weather. Somewhere in the history of Venus, something went very wrong, however, and today its surface is hot enough to melt lead. While Earth can never become Venus, the latter has much to teach us—particularly about greenhouse atmospheres, the reason for its high surface temperature.[30] Unlike Earth, Venus never had humans stomping around building factories there (that we know of). Which means to study that planet—specifically, to compare it to Earth and monitor its evolution alongside ours—is to study Earth, eliminating the human variable.

Likewise, when the rover Curiosity observed the sand dunes of Mars, planetary scientists noticed formations that do not exist on Earth.[31] It was a small finding, but a good example of a then-blank spot on the map for geological processes and how worlds work. Into the model it goes. It's not just robotic explorers, though. Once humans live and work on Mars, they may be able to access that planet's pristine climate record, just as glaciologists take deep coring samples on Earth. This will be vital from a comparative planetology standpoint and yield insights as-yet unimaginable, or at least: impossible as-yet to prove. Atmospheres also play a role in Martian studies as they relate to Earth. If Venus is hell as depicted in Renaissance paintings, Mars is hell as described in Dante, namely: freezing. Venus has a thick atmosphere, and Mars has a wispy one.

In the distant future, there is one area where the very fate of humanity itself rests on a robust space exploration program. Specifically: The planet Earth has an expiration date.

The End of All Things might be caused by an irresponsible humankind with respect to the environment, but more likely, Earth might well be doomed to some unsuspecting fate. We are one giant asteroid away from losing all the dreams of our ancestors. Populating the stars—not in the way that the International Space Station is supported by Earth, but in a meaningful, self-sustaining way—is the only future where humanity is spared oblivion. This is the apotheosis of sustainability.

But in the more immediate, robots rolling across alien plains in our solar system, and artificial moons circling their skies, are not the only places where research in space pays dividends on Earth. The International Space Station is a global laboratory. The governments of the world encourage scientists of their countries to put forth experiments for astronauts to perform.

Sometimes, the research concerns space exploration itself—for example, studying the way prolonged exposure to microgravity affects the human body—but just as often, the experiments seek to understand how new medicines, developed in the unique environs of space, might be more effective than those developed on Earth. It is not just the physical sciences, though. The denizens of our unusual laboratory in orbit are themselves studied by sociologists and psychologists. From an engineering standpoint, meanwhile, the things that keep the Space Station operating also have direct applications on Earth. Solar panel technology is one such example of that, though everything from the miniaturization of computers to advances in optics and robotics are in a constant state of advancement because of the work by astronauts in Earth's orbit.

A Conversation with

Paolo Nespoli

PAOLO NESPOLI has been an astronaut of the European Space Agency (ESA). He worked for the European Space Agency from 1991 to 2018 as an engineer, and also as an astronaut from 1998 onwards. He flew in space three times: in 2007 on a Space Shuttle mission building the International Space Station; then, he flew twice on long-duration missions on the same station in 2011 and 2017. He has been in space for a total of 313 days.

When did your interest in space, and space science, start? How has your journey to becoming an astronaut been?

The idea of becoming an astronaut grew in me when I was a kid, while I was watching the Soviet Union and the United States' space missions aimed at conquering the Moon. It took me quite some time to find the courage to pursue that dream.

During my study years, I actually figured out that I like science, I like engineering and I developed this idea of working in the space field, and that is why I applied to be part of the European Space Agency. I worked there for several years as an engineer, while applying to various calls for astronauts and finally I was selected in 1998.

What does it mean to do research in space?

Making research in space is a multifaceted concept. While we are now focusing on space as an opportunity for research, in fact it is at the same time an opportunity for research, for commercial activities, and for exploration. Although these are different activities, their boundaries may be blurred. For example, when discussing research, one might immediately think about the International Space Station. However, future missions to Mars might provide scientists with a huge amount of valuable information as well, but could not the latter be considered as exploration as well? So, when elaborating on space research, I think it is important to consider its multifaceted nature, otherwise a lot of opportunities might be missed. Another meaningful example concerns space tourism. Might it be considered as research? While to some extent it might be labeled as vacation, actually I believe it may represent a good opportunity for humanity to advance its knowledge and capabilities. People usually tend to forget that we are guests on Earth and do not realize that we cannot do whatever we like to this planet. Looking at it from space can change our perspectives, making people aware that they can (negatively for us) influence Earth living conditions and making them willing to do things differently. If we do not understand this, we are not going far as a species, thus having tourists who go in space and appreciate this might be very important. So, is space tourism not relevant because it does not correspond to our canonical idea of research?

How would you say research in space can benefit terrestrial activities? Which are the fields that can benefit more from it? Why should we invest in it?

Space provides us with conditions that cannot be found on Earth, which means that, in space, we can do things that are totally impossible on Earth. For example, on the Space Station orbiting around Earth you experience a condition of microgravity: when you remove gravity, which is a very strong force, other forces become apparent and you can measure them and understand how they work. It is not that they do not exist on the ground, but they are completely squashed by gravity, making it absolutely necessary to remove its influence. Obviously, this does not concern only objects but humans as well. One has to think that those who go to space are human beings, and their body feels the microgravity environment and starts changing. Somehow, they become "extraterrestrials"; this change is measurable and can provide important data, for example, to health studies here on Earth.

Another thing that comes to my mind is the living conditions: on the International Space Station, you constantly live in a team and in small environments, working together, in a place that is full of equipment. You need to learn how to behave in these conditions. One could say that even sociologists can get more data from space activities.

Is it correct, then, to say that you basically collect data in space, instead of providing new findings?

A lot of people call astronauts scientists. In fact, I do not think that astronauts are scientists in space. While it is true that they conduct research, still they do not conduct research that they have invented, but somebody else's experiments, and sometimes it is not even requested that they understand everything that they are doing. Astronauts have to understand how to collect data and how to do it properly. But they are not there to draw conclusions, they are there to collect data.

Usually, astronauts' teams are composed of people coming from different countries. What is the connection between space research activities and international collaboration? Is the space sector different from other fields in this respect?

Cooperating means that I share with you what I know, I learn from you what I might not know, and together we reach a result. When some nations have an enemy on the ground, for any reason, they hide their own knowledge and they do not want to give it away for free.

This holds in general. In space, cooperation reaches a different level somehow. In space, it looks like we are able to forget about our differences on the ground. Maybe because when we are out there, we look at this endeavor as an objective that has to do with humanity. It is like we try to find the optimum for humanity as a whole and not for a single country. On the International Space Station, astronauts actually work side by side even though their nations may not be really good friends on Earth, so we can say that space is a factor that alters regularity, in this respect.

In the near future, we will probably see private space stations in Earth orbit. Do you think that private initiatives will change the way in which research is conducted in space?

Commercial activities are only viable if they produce returns, and this is not always compatible with research. While researchers' primary goal is exploring the unknown, taking findings and making them commercially valuable is someone else's task. To be fully honest, I would suggest to every company to invest at least part of their resources to unconstrained research as great benefits might derive from unplanned investigations. Was Christopher Columbus expecting to land in the Americas and to discover a new world? Discovering implies being open to something that we cannot predict. Obviously, some findings eventually prove to be useless, but some others turn out to be innovative, or even disruptive. Thus, governments should adopt a two-sided approach. On one hand, they need to enhance scientists' capabilities and keep providing them with opportunities to do research. On the other hand, they should encourage commercial entities to take research findings and make them commercially viable, providing innovative value to communities. Once the International Space Station will be over, I wish that humanity will not give up the opportunity to conduct unconstrained research in space, namely to have a laboratory in microgravity allowing us to do all sorts of crazy things that we might only imagine on Earth. If space stations were the prerogative of private entities', they would probably impose stringent constraints to research activities due to stations' high costs. That is when governments should step in through international cooperation initiatives, acknowledging that space research should be preserved to benefit everybody.

Why should governments cooperate in this field?

Why did I say governments? Because I realize that preserving research is a burdensome duty and one government cannot or would have difficulties in sustaining this effort all by itself. This is why we need to go back to the paradigm of building international cooperation, and, most fundamentally, it should be one where no single nation establishes itself as a leader. In fact, when an international project is led by one nation, it becomes a project of that nation with all the others following. In these cases, if something happens to the leader, the whole project is at risk. Today, a clear example is offered by the International Space Station. While the United States and Russia are using the Station's future as a pawn in the context of their power struggle, the European Space Agency, although participating, cannot influence the discussion. We should aim, instead, to a more democratic international cooperation, to let more balanced geopolitical *equilibria* emerge, to guarantee to research activities a stable environment, and to let as many countries as possible benefit from research conducted in space.

What do you think about the Artemis program? It has an American leadership, but it collects competences and support from many other countries.

Let me underline what is essential, the Artemis program is a United States' initiative that aims at bringing an American woman and an American man on the Moon, not "generic" human beings. Only the United States can decide whether some people from different nations will be able to join them. This is why the Artemis program is not an international project. Surely the United States is allowing other countries to contribute, but it is not as they would not be able to achieve their purpose by themselves. Real cooperation does not concern "participating" in a project but rather being fundamental contributors and decision makers. This is again similar to the case of the International Space Station. While it is called international, if the United States would cease to support it, it could not survive, while the same would not be true for other nations.

Could other nations actually play a more influential role?

I hope so. I wish that other nations could develop new competencies and find additional resources to enhance their positioning within international programs, and that the United States would let others step in. However, I am aware that my hopes are difficult to realize. First, the United States has no valuable reasons to give up their leading role. Secondly, other nations do not always have enough resources and technical knowledge to make their contribution crucial. This is where politics should intervene to create a more equitable balance, and this is not characteristic to the space sector only, but to many other strategic fields, such as the military industry or the transportation and aeronautical ones. At the same time, space uniqueness might help here. When we are outside of Earth altogether, we feel more united, we feel more human, forgetting the flag we bring on our shoulder. In space we work together to achieve the same goal for the sake of humanity as a whole. We should take care of such a realization and amplify it to as many human activities as possible.

Satellite for Char

Services

Climate

ge

Every living creature changes the Earth in some way because we are all of the Earth. As a species, humanity, whether intentionally or otherwise, can and has altered fundamental aspects of the planet. This is a testament to the achievements of humankind.

To lament the automobile, electricity, or indoor plumbing is to slip into environmentalist self-parody. Conversely, to disavow our responsibility to clean up our messes, whether air, sea, land, or orbit, is the height of irresponsibility and a self-defeating betrayal of the planet whose resources yielded our progress.

Environmentalism as a political issue is not particularly controversial. In the United States, for example, as far back as the 1980s, there has never been a public opinion poll taken by the firm Gallup in which a majority of Americans did not express significant concern over issues as diverse as water pollution to the loss of tropical rainforests.[32] Since 2014, Americans have consistently ranked environmental protection over economic growth. To be sure, American elections have not always yielded officials consistent with the stated beliefs of its people, but even accounting for the so-called "Bradley effect", which posits that voters give pollsters answers in accordance with "social desirability"—that is, they give answers they believe to be publicly acceptable, rather than honestly—it still arguably reveals that people know on some animal or social level that a clean environment is the "correct thing".[33]

The attainment of clean air and water is an absurdly obvious, high priority goal on which everyone alive should agree. The divergence is in how best to achieve that goal, which brings us back to abolishing the automobile.

One visit to Los Angeles would tell you that such a policy would surely do wonders for air quality. Sweeping, civilization-wide changes simply seem not in the offing, however, which means if there is to be progress, it will be incremental. This is a terrible dilemma for those who believe Earth is in a race against time, where an environmental tipping point is only years away, and that by the time the increments add up to real change, it will be too late.

Any meaningful change, dramatic or subtle, and any hope of persuading those most resistant to climate solutions, can only come through the power of data, which is where space sustainability again enters the picture.

To understand the state of the Earth and trends in its health, scientists must monitor it continuously, establishing an ever-updated chart of vital signs for the planetary patient. With such information, accurate models might be written and informed decisions might be made. Humans have been pumping into the air dreadful byproducts of industrialization since the 1800s. Now more than ever, however, we can measure their effects with high fidelity, and perhaps find ways of mitigating the worst damage to our home planet.

Spacecraft in Earth's orbit can collect that data. Persistent monitoring reveals snapshots of any given moment, and how the climate is changing over time. Data collected, for example, from NASA imagery over Antarctica can be correlated with temperature data in the tropics and levels of emitted carbon dioxide around the globe.

According to the World Meteorological Association, there are fifty-four "essential climate variables" that must be monitored so that planetary scientists can properly assess how the world is changing. Of these, at least half must be measured from space.[34] Among the measurements necessarily taken by satellites are surface wind characteristics, including how hard they blow and in what direction; the concentration of water vapor throughout the atmosphere; rain levels; and the nature of clouds.

Satellites can also track greenhouse gasses like methane and carbon dioxide. To understand the oceans, spacecraft can measure the temperature of seas at the surface, as well as water levels, the expanse of sea ice in the cryosphere, and something as simple as the color of the ocean, which reveals sediments, organic matter, chlorophyll, and phytoplankton. In addition, space-based platforms allow both government and industry to do simple reconnaissance when choosing prime locales for power plants. When natural disasters strike, imaging spacecraft and telecommunications satellites give communities the means to survive. As an engineering matter, as far back as the 1950s, the development of solar power for spacecraft unlocked its utility back on Earth.

Beyond Earth science spacecraft, satellite products include telecommunications systems and navigation services such as the Global Positioning System (GPS).

Though the backbone of the Internet consists of cables laid at the bottom of the world's oceans, orbital satellites are transforming communications. They solve the "last mile problem", in which major land-based telecommunications companies have proven unable to get broadband into people's homes. As the Starlink network demonstrated when Russia switched off Ukrainian access to the Internet, space allows people in repressed or conflicting areas to remain connected to the rest of the world. Moreover, GPS systems go far beyond helping commuters avoid traffic congestion. Today, farmers rely on navigation systems for "precision agriculture", enabling twenty-four, seven farming processes with maximum efficiency of time and harvest yield.

Such extraordinary abilities reflect the great technological leaps in spacecraft technology since the launch of Sputnik in 1957. Satellites are smaller, and it is increasingly common that their services are provided by way of constellations rather than single, capable spacecraft. Those Earth observation satellites circling the Earth can collect higher resolution imagery of our planet, faster than ever before, which benefits farmers and war planners alike.

Data latency from sky to ground is lower than ever, improving communication services. And perhaps most importantly from a sustainability perspective, engineers now incorporate collision-avoidance technology in their spacecraft.

The only way to achieve the Sustainable Development Goals (SDGs) as set forth by the United Nations is to incorporate space sustainability, for the reasons stated previously. There is no other way to view global infrastructure, or the Earth's myriad ecosystems.

The very first goal is to end poverty, which requires, among other things, the ability to monitor natural disasters and connect responders in the stricken area as well as experts on the other side of the planet. Something as simple as imagery of the ground can enable response vehicles to travel safely to isolated areas.

Much work remains in fully developing and leveraging satellite capabilities, but private players in the satellite market have expanded the possible services available, and state agencies are leveraging their research centers as well as treasuries to kickstart private initiatives. Even the simple act of raising awareness benefits space sustainability, and by extension, the human race.

It might be a cliché to say "knowledge is power", but despite our extraordinary advances in orbit, there is still so much we do not know and cannot yet know. In time, and by working responsibly in orbit, we soon will.

A Conversation with

Simonetta
Di Pippo

SIMONETTA DI PIPPO is Professor of Practice of Space Economy at SDA Bocconi School of Management, where she is also Director of the Space Economy Evolution (SEE) Lab. She is an Italian astrophysicist, and before joining SDA Bocconi, served for eight years as Director of the United Nations Office for Outer Space Affairs (UNOOSA) in Vienna. Prior to being part of UNOOSA, she was Director of Human Spaceflight at the European Space Agency (ESA), Director of the Observation of the Universe at the Italian Space Agency (ASI), and led the European Space Policy Observatory at ASI-Brussels.

Could you share with us when and how your sensitivity to satellite-based solutions to climate change and to how to spread their adoption was born?

I would say that the time I have spent at UNOOSA has been key to bringing forth my sensitivity. Before joining UNOOSA, I spent many years working on outer space missions, not dealing with their benefits for Earth and collaborating mostly with space-faring nations that, by that time, were apparently less exposed to climate change issues. As a result, I had never been challenged to employ satellite data and infrastructure to tackle climate change related issues. On the contrary, throughout my experience at the United Nations, I have extensively engaged with a much wider range of nations, which required me to change my perspective and to consider how space technologies can contribute to their development. In that context, I got the opportunity to observe the urgency of climate change challenges while witnessing developing countries' unpreparedness, and thus to start nurturing my sensitivity to the topic.

What do we mean by satellite services and applications?

The expression "satellite services" is meant to indicate Earth observation (EO), satellite telecommunications, and navigation services offered by satellites positioned in Earth's orbits. EO refers to the collection of information about the physical, chemical and biological systems on Earth, usually via imaging devices. It allows monitoring of natural events, as well as anthropic activities, offering valuable data for descriptive, predictive and prescriptive analytics. Notably, weather forecasting activities are performed by a subgroup of EO satellites. The term "satellite communications" identifies the use of wireless communication satellites that, through a transponder, receive and send data from and back to Earth. Nowadays, satellite communications are crucial to strengthen land-based connectivity and to connect remote locations that are not covered by terrestrial networks. Last but not least, navigation satellites provide autonomous geo positioning services. To simplify, navigation devices on Earth exploit different satellite signals to determine their own location. On the other hand, "satellite applications" are satellite-based products or services tailored for specific uses and, as such, bundled with other technologies and/or services.

How is satellite technology evolving?

Since the first satellite, Sputnik 1, was launched by the Soviet Union in 1957, satellite technologies have constantly advanced, increasing their services' quality. In EO, this translates into higher images' resolution and frequency; in communications, into lower latency and broader coverage; in navigation, into greater trustworthiness and, as new competitors are entering the market, interoperability. Another relevant trend, today, is represented by satellites miniaturization, which, together with the need for higher coverage and images' resolution, has concurred to the development of satellite constellations, namely groups of up to thousands of satellites, usually positioned in LEO, that work together as a single system, rising services performances. Finally, the increasing menace of accidents between space objects (due to orbits crowding) is pushing satellite manufacturers to design new collision-avoidance and satellite-recovery systems.

What Sustainable Development Goals (SDGs) can benefit from satellite services and applications?

All of them! EO, communications and navigation services are not only useful but key to achieve all the SDGs. Just think about EO potential for monitoring natural ecosystems or infrastructure planning, building and maintenance, satellite communication potential for connecting remote communities, improving education and healthcare, as well as navigation services' role during crisis management. And these are only a few examples. Without space, we could not survive in our current social landscape.

How can satellite services and applications contribute to SDG number 13, "Take Urgent Action to Combat Climate Change and its Impacts"?

The opportunity of monitoring Earth from space helps humans get a deeper understanding of nature functioning, monitor how climate is changing over time, model future scenarios and design realistic and effective mitigation and adaptation policies. Concerning the energy transition, information obtained from space allows finding the best possible location for siting energy plants and remotely controlling construction sites' progresses, optimizing plants' performances through weather forecasting and better balancing the energy mix, as well as performing predictive maintenance to reduce supply interruptions due to unexpected damages. In combination with EO, satellite communications, geolocation and navigation are fundamental in case of natural calamities, playing a key role in the entire disaster management cycle, including prevention, preparedness, early warning, emergency response and downstream reconstruction of cata-strophic events.

Notably, technical advancements in satellite technologies also provide sev-eral spin-offs that contribute to enable new solutions for climate action on Earth. Do not forget that photovoltaic systems are among these. In 1958, the Vanguard 1 satellite showed off solar panels, which powered all its ac-tivities. Thanks to those solar cells, not only the mission was extended to 1964, well after original expectations, but the basis for the photovoltaic plants that we are currently using on Earth were set.

Who is currently adopting satellite-based solutions for climate action?

Public institutions employ them to set their agendas and to monitor their pol-icies' impacts and target actors' compliance, aiming to recursively improve their decisions. At the same time, they also benefit from satellite services for natural disasters management, to make interventions timelier, more efficient, and less hazardous for emergency operators. Private companies adopt satellite services to optimize their processes with the aim of reduc-ing their environmental footprint (just think about precision agriculture or route optimization in logistics activities). Non-governmental organizations can adopt satellite service-based solutions for similar purposes. Notably, EO is also particularly relevant as it provides valuable material for awareness campaigns, offering unmatched perspectives on our planet's fragilities.

While satellite services have long been dominated by public players, private actors are assuming an increasingly relevant role in the mar-ket. How has the offering of satellite services and applications for climate action evolved correspondingly?

Private players' entrance into the market has multiplied the number of available products and services, making them more fit to a commercial demand. Given satellites' strategic role for public institutions' activities and agendas, they have long been designed to respond to public institutions' needs. However, while this scope is still served, the necessity of new prod-ucts and services has rapidly emerged as soon as the industry has opened to commercial operators. Companies from non-space industries are asking for applications specifically tailored to their needs and while some steps have been already taken in this direction, there is still a long way ahead. A strict dialogue between supply and demand actors, education and informa-tion campaigns, cross-industry collaborations, countless proof-of-concepts and the rise of new companies specialized in producing industry-specific applications are all factors that might contribute to the ongoing development of satellite applications offerings.

At global level, does a data-accessibility issue exist with respect to satellite data for climate action? Who is responsible for addressing it and what are the main solutions currently available?

Non-space-faring nations have only indirect access to satellite data and only some of them can acquire information from private suppliers. Nonetheless, current global challenges, such as climate change, require coordinated actions from all countries around the globe, implying the necessity that all Nations possess the technological means to join climate-related international agendas and strategies. As a result, several initiatives have emerged, which gather the international community, including both public and private players, to make key satellite data available for all countries. At the same time, some public institutions are adopting free-open-access policies to make their satellite data available for anyone who is interested in them, supporting the rise of uncontrolled entrepreneurial solutions to climate change-related challenges. In my tenure as co-chair of the World Economic Forum Global Future Council on Space, I led the preparation of an important white paper, Space for Net Zero, which, among other proposals, suggests the creation of an Earth Operation Centre federating all the available space date to help better prepare humanity to tackle the consequences of the current climate crisis.

What are the key steps that you envision to enhance satellite services and applications' contribution to climate change mitigation?

There are at least two key pre-conditions to deal with any problem that concerns the entire international community. At first, a common awareness of the phenomenon. If different stakeholders have different perceptions about the issue, finding a shared solution might be much more challenging. Therefore, outlining and conveying an effective narration of the problem, capable of touching everyone's sensitivity and, even more, interests, is fundamental to involve both already committed and more skeptical actors. Unfortunately, today the opportunity to effectively pursue sustainable development is threatened by a narration focused on mitigation and adaptation strategies. On the contrary, the global community should strongly react to scientists' demand to reduce human pressure on natural ecosystems joining efforts to achieve "net-zero" (the balance between the emitted and removed green-house gas emission) or more (removing more emissions than those emit-ted). Secondly, everybody needs to be autonomous within the cooperation framework. To put it another way, all the stakeholders should be competent enough to discuss with the other involved players on an equal footing. A virtuous cycle is triggered only once everyone can join international collabo-ration initiatives and platforms. Today, the United Nations are the sole insti-tution that can encourage and support this global effort, engaging regional and local partners. It needs to rapidly increase its relevance on the topic as the climate crisis is rapidly evolving and Earth ecosystems react according to only partially known dynamics. In this context, space technologies are key as they enable decision-makers to better understand the challenge that we have before us and how to better respond to it.

Spa
Solar

ce
Power

The scarcity of energy has for centuries limited the growth of developing nations around the world; caused damage to the environment as humans search for new ways to increase supply; sparked wars over previous resources and the fortunes they might command; and driven innovation in the ways power is generated and consumed.

Clean power on the ground today includes wind turbines, solar panels, hydroelectric dams, biofuels, geothermal techniques, hydrogen cells, and nuclear power. Each has its drawbacks.

Wind power can only be placed in certain regions, is unreliable (the wind must blow), and can harm local avian populations. Solar power demands massive footprints on the ground, sometimes necessitating deforestation, displacement of people, and disruption to local ecosystems. Moreover, it is only sunny half the time. Hydroelectric power is devastating to local ecosystems, flooding regions upstream of dams, and parching regions downstream. The increasingly dry seasons concomitant with a warming globe also result in less power production. Biofuels require, as the name suggests, biology, which means water (a precious resource in an increasing number of areas) and harms biodiversity. The drilling and water injection necessary to generate geothermal power increases earthquake risks, and disrupts natural geysers. For hydrogen fuel cells, the namesake element can leak, the cells can explode, and the energy content per volume unit is poor. Nuclear power has an image problem in the United States and in Europe in particular and nuclear research is banned outright in many nations. The Fukushima disaster in 2011 did not help the cause for a wider embrace of nuclear power.

The downsides of each of these power sources, of course, pale in comparison with the damage wrought by the burning of fossil fuels, the repeated and ongoing catastrophes caused by drilling for oil and natural gas, and the destruction

of the natural world that comes from coal extraction. This is to say nothing of the inevitable depletion of these non-renewable resources.

Practically any multi-pronged clean energy solution is preferable from a sustainability standpoint to the continued incineration of fossil fuels, but those solutions lack the existing infrastructure of "dirty power", and aside from nuclear power, do not tend to scale well across countries the size of the United States or China in a manner palatable to taxpayers or governments. Short of a revolution in efficiency and price, it is simply hard to foresee the wholesale replacement of power generation the world over.

Just such a revolution, however, might be possible in orbit. Space solar power (SSP) solves each of the negatives of terrestrial solar power.

The technology, which involves placing solar arrays in orbit, takes up no footprint that might lead to deforestation or displaced people or wildlife. Better still, if placed in a proper orbit, the necessary collector arrays are in constant, unfiltered sunlight.
Historically, the downside to space-based solar power has been cost. In the days of expendable rockets, lifting anything heavy to orbit was expensive on the order of absurdity. Fortunately, we no longer live in that age. In addition, there has been the problem of getting the collected power from space back down to Earth. The most feasible, posed solution has been to convert the Sun's collected energy into microwave energy, which can be beamed to downlink stations on Earth.
The obvious benefits of this technology have not been lost on global powers, and nations including the United States, China, and the United Kingdom are funding research into ways space-based solar power might be deployed.[35] Reportedly, China has invested heavily in the technology, with plans to deploy it in force by 2035.[36] Meanwhile, in 2020, the United States Naval Research Laboratory generated solar power in a satellite and converted it to microwave energy.[37]

Ironically, one of the key threats to the potential, almost magical solution to the world's energy needs is another major space sustainability issue: space debris.

However, should that solution be implemented on a meaningful scale, it could conceivably generate two gigawatts per day of power to major markets. This is analogous to the peak output of Hoover Dam.[30] Unlike Hoover Dam, however, a space-based solar collection system would not be limited to the shifting templates of nature.

To wit, dam-based power is inconsistent. Even Hoover Dam of late is only putting out one-quarter of its potential power.[39] Given the diminishing water supplies at Lake Mead, this problem is unlikely to improve. Just as water is no concern for space-based solar power, neither is cloud cover—the bane of planners depending on power for photovoltaic cells. No sunlight, no electricity for communities. Renewable energy efficiency is measured in megawatts per square-meter. Space solar power involves two chief systems: the solar array satellite in Earth's orbit, and the ground station receiving collected power.

The aforementioned two gigawatts space-based solar power system would need to be about 30,000,000 square-meters large, with an equally large receiver on Earth. The power delivery would amount to 48 million kilowatt-hours of electricity per day, year-round.

A terrestrial solar farm of the same size—30,000,000 square-meters—would produce half that power in summertime, and a sixth of the power in winter. On an overcast day, the power generation could be nil. (Note that 30 square-kilometers is not outside the bounds of imagination. Indeed, such a solar park or solar receiver on Earth would be smaller than Pavagada Solar Park in Karnataka, India.)

Even modest space solar power projects could be useful for remote outposts, such as mining operations. And not just mining operations on Earth.

While wind farms and hydroelectric dams would work poorly on the Moon, space solar power would work very well.
Indeed, it is the case where developing power sources for a sustainable spacefaring future would yield a power source for the near-term on Earth.

Conversely, developing this sustainable power source for Earth would enable a spacefaring future. Humankind's ambitions for exploration, and NASA's and ESA's plans to establish a lunar outpost as a precursor for eventual Mars settlement, might well hinge on space-based solar power being implemented.
In many ways, this is evidence that sustainability goals need not be distractions or impediments to interplanetary endeavors, but rather, are emblematic of the integrated nature of space sustainability.

Ultimately, while space solar power is ideally suited for multiplanetary environments and vastly enables the making of humans into multiplanetary species, it remains the activities on Earth that can threaten its future employment.

Specifically, the transmission technologies that would send power back to Earth from space could indeed be weaponized.
If, however, the better angels of our nature prevail, cooperation can bring a quicker close to our reliance on fossil fuels for power generation. Wireless power transmission from space to ground stations will require a dedicated part of the radio spectrum. International cooperation would allow a space-based array to direct power to multiple countries. Likewise, technology transfer

between nation-states will require cooperation. Here, previous space exploration has established a working precedent: particularly those agreements that facilitated, launched, and keep in orbit the International Space Station.

A Conversation with

John C. Mankins

JOHN C. MANKINS is an independent consultant, entrepre-
neur and professor living in California. He worked at NASA's
Jet Propulsion Laboratory, working for CalTech, for ten years, and at NASA Headquarters for fifteen. He is currently
a board member of the National Space Society; an advisor
to the Beyond Earth Institute; vice president of the Moon
Village Association; co-chair of the Power Committee of
the International Astronautical Federation (IAF) and of the
Permanent Committee on Space Solar Power of the International Academy of Astronautics (IAA).

What is Space Solar Power (SSP)?

SSP is the idea of placing in space a platform designed to harvest incoming sunlight and to convert its energy into electricity. The collected electricity could then be used to generate electromagnetic (EM) waves that can be used to wirelessly transmit power to a remote location—for example from space to receivers on Earth. If the "solar power satellite" platform is placed in a high Earth orbit (such as the geostationary Earth orbit, GEO, at 35,786 kilometers), where the Sun shines 99.95% of the time, the platform can deliver solar energy wirelessly to terrestrial markets approximately twenty-four hours per day, seven days per week.

What kind of space and terrestrial infrastructure are necessary for SSP to work?

Due to the large number of its components, SSP involves exceptionally low-cost Earth-to-orbit (ETO) launch systems and related infrastructure at launch sites. Other needed infrastructure includes in-space transportation systems, in-space assembly and construction systems, orbital traffic management systems, command and control systems, and others. However, "upstream" of these elements of infrastructure, large-scale manufacturing facilities will be needed since the making of solar power satellites and ground receivers will be a tremendous industrial undertaking—analogous to today's personal electronics industry.

What is the state of the art of the technology and when do you expect it to be demonstrated and available?

All the physics and basic devices needed for SSP have actually been well established. However, there are various ways that SSP might be implemented—some of these are more technically challenging and others are less so. One highly promising approach is to use highly modular space systems architectures such as those involved in the so-called "SPS-ALPHA" (which stands for solar power satellite by means of arbitrarily large phased array), studied by NASA's innovative advanced concepts (NIAC) program in 2011. Using a "hyper-modular" approach would indeed ease transportation and assembling activities. As a result, SSP could be developed and demonstrated at its systems level on Earth within three years, in low Earth orbit within five years, and in space (middle Earth orbit, MEO, or GEO) within ten to twelve years.

Which actors are currently working on SSP at global level?

As of 2022, more countries and organizations are working on SSP and wireless power transmission than ever before. Leaders in conducting research and development activities include, for example, Japan which has been pursuing SSP research and development for more than thirty years and has its own plans for a near-term SPS demonstration. Both the government, through the Japan Aerospace Exploration Agency and the Japan Space Systems, and non-governmental players, such as, for instance, the University of Kyoto and Mitsubishi Corporation, are currently working on this. In the United States, the Department of Defence Air Force Research Laboratory is contracting SSP-related projects to Northrop Grumman Corporation. At the same time, Caltech, supported by a large grant from a Californian billionaire, and a number of small start-up firms are also developing the SSP concept. China is also interested in exploring SSP-related opportunities and is doing so through the China Academy for Space Technology and various universities. The United Kingdom (UK) has recently announced funding to start SSP research and development activities shortly. The UK Space Agency, Department for Business, Energy & Industrial Strategy, and the non-governmental "Space Energy Initiative" are all active on this. Finally, while the European Union has no current plans vis-à-vis SSP, the European Space Agency has completed preliminary studies of SSP and has now proposed a new program—tentatively called "Solaris"—conduct R&D directed toward a future European solar power satellite.

For which applications (or in which situations) can this technology be used?

SSP can be used for a variety of applications, including both terrestrial and outer space activities. Looking at Earth, to start, a single large solar power satellite could deliver perhaps 2 GW to a major baseload market, such as a mega-city (e.g. Chicago, Singapore, Kinshasa, London). Due to its constant exposure to sunlight, SSP can be especially valuable when weather does not support ground-based renewable energy (e.g. when it is overcast, and ground photovoltaic arrays are not producing electricity). Moreover, at a moderate scale—perhaps delivering 100 MW—it can be very useful in providing power to a remote location, such as mining operations. Concerning outer space uses, SSP will be very important for a wide range of ambitious applications, including operations on the Moon, power for resources development at small bodies in Earth's vicinity, and for use in the Mars system.

What kind of challenges do you envisage in the development, deployment, and use of SSP?

A central challenge for SSP is surely that of transportation of the modules from Earth to their desired operational location—typically in or near GEO. In fact, as space transportation costs are strictly related to payloads' weight, delivering large infrastructure as that of SSP still entails high expenses. Unless new approaches, this will remain the major obstacle. Similarly, while all of the physics for, and the basic component technologies involved in SSP are already available—and have been so for some decades—new materials and new devices are constantly emerging that might optimize the SSP concept. These must be applied and matured to be used in SSP systems and supporting infrastructures. Finally, one or more portions of the electromagnetic spectrum must be formally identified for use in accomplishing wireless power transmission. However, to tackle all SSP-related challenges, I would say that, more than anything, new mindsets are needed, namely new ways of employing existing and novel technologies to the problems of SSP.

How do you think that the mentioned challenges can be solved?

In the longer term, the use of lunar resources is one very promising opportunity for reducing the costs and environmental impacts (e.g. emissions related to launches) of transportation. The Moon's surface regolith (the lunar soil material) comprises many of the elements out of which the satellite may be manufactured, including aluminum, silicon, iron and others. If system components (perhaps structural systems) can be manufactured on the Moon, then these could be transported to GEO at very low costs. Such components could then be readily integrated via robotic construction systems with other modules delivered from Earth, for example.

Is energy production from SSP uninterrupted regardless of the solar power satellite position?

The power generated by a solar power satellite depends on the orbit in which it operates. In the case of GEO, where SSP should be located, sunlight is available approximately 99.95% of the year. Only during a couple of weeks around the Spring and Autumnal equinoxes—when GEO passes through Earth's shadow—power generation is suppressed, and then only for a maximum of about one hour at local midnight.

What is the end-to-end efficiency of space solar power from its initial production in space to its distribution to electrical grids on Earth?

The most important performance parameters to consider when evaluating novel sources of electricity such as SSP include the cost per kW to deploy the system, the cost per kW-hour for the electricity produced, and the availability of the power. In addition, the end-to-end efficiency with which a technology such as SSP converts energy from one form to another directly impacts the total mass of hardware required to produce a given amount of electricity. The end-to-end conversion efficiency, with off-the-shelf components, would likely be about 10-12%; with advanced components now in the lab that number would likely improve to 15%-20% or more.

Let's compare MW/meters-squared produced by SSP and different forms of low-carbon energy (e.g. nuclear or renewables).

In summary, an SPS-ALPHA would deliver about 1 MW for each 15,000 square-meters, and about 1 MWh daily for each 6,250 square-meters. Let's compare this to three existing renewable energy options: ground solar power, hydroelectric power and wind energy.

First, by way of comparison consider a notional terrestrial PV array with the same area as the solar power satellite receiver—30 square-kilometers, i.e., 30,000,000 square-meters. This array will produce in the middle of summer a peak power of about 6,000 MW and energy over 24 hours of about 24 million kW-hours. However, that same photovoltaic system will produce only 2,000 MW power on average in winter, and only 8 million kWh of electricity on a winter day. And, of course on an overcast day, the energy delivered by the ground photovoltaic system will be extremely small. Hence, a ground solar power system will produce in winter a daily average of 1 MW per 90,000 square-meters.

Second, consider a hydroelectric power plant such as Hoover Dam in the United States. This plant has a rated capacity of about 2 GW peak capacity, however, because of the variations in rainfall over the year, it produced on average only about 500 MW (0.5 GW). This works out to some 12,000 kW-hours per day. Considering the lake's extension, this equals 1 MW (average) on an area of 1,280,000 square-meters.

Finally, the land area required for wind turbines is approximately 1 MW per 345,000 square-meters. As can be seen, the land area required for SSP is a small fraction of all other typically mentioned renewable energy options.

What technological advances might make SSP economically competitive with renewable energy alternatives?

Although modular space solar power concepts can be implemented with existing technologies, there are a number of needed advancements that will make highly modular SSP much more competitive with terrestrial renewable energy alternatives. Additional infrastructure, system and technology advancements include (1) exceptionally low-cost launch systems at less than 100 dollars per kg to low Earth orbit; (2) highly-efficient photovoltaic arrays at greater than 35% conversion efficiency (sunlight to electricity); (3) autonomous robotic assembly and construction systems; (4) low-mass structural systems; and others.

Who should fund SSP?

I believe that both government agencies and private sector entities must be involved in realizing SSP. More specifically, during the research and development phase, increased levels of national government support "blended" with private sector funding will be needed to reduce later financial risks perceived by potential non-governmental investors (such as venture capital funds at first and later debt-financing). However, the deployment and operation of mature solar power satellites will likely involve almost complete funding by the private sector, potentially in combination with local and regional actors.

Would it be possible to integrate the supply of energy from space into current power markets? Do you see any criticalities?

It will be not only possible, but actually essential for SSP to be integrated with terrestrial sources of energy. SSP may play a key role in allowing very large percentages of the electricity supply to be sourced from renewable, but intermittent technologies such as ground solar, wind, or hydro, while integrating to accommodate their periodic "downtimes", as, for example, during night-time hours or during periods of extended overcast. By doing so, the need for substantial energy storage and oversizing of ground renewable sources, which are currently the major obstacles to renewable energies overriding fossil fuels, can be reduced or eliminated.

Is there any legal framework for the SSP? Are there regulations that can be borrowed from terrestrial sectors?

SSP does not have a dedicated regulator framework. There are rules and regulations that have to do with the isolating and operating substations that handle the power generated by conventional sources; certainly, these may be entirely relevant to the operation of SSP systems. In addition, there are rules in various countries that concern exposure to microwave frequencies which also might be taken into consideration to regulate SSP security aspects.

Do you think that the implementation of the regulatory framework will speed up or slow-down the SSP development and use?

The implementation of a regulatory framework will definitely speed-up SSP development and use. Initially, without the assignment of a specific part of the radio spectrum to use for wireless power transmission, SSP systems cannot be deployed and operated. In addition, in order to enable a single solar power satellite platform to serve multiple countries, the receivers in those countries must be fabricated to receive and convert the same specific frequency being transmitted by the same solar power satellite. Realizing such standards for international wireless power transmission will require a clear regulatory framework.

It is a common opinion that investing public funding in space should have a low relevance in the agenda of governments as multiple economic, social, and environmental issues on Earth need to be solved first. What is your opinion about this statement? Can you elaborate on the benefits that space, and specifically the SSP, generates to society? What Sustainable Development Goals (SDGs) can be addressed by the SSP?

Future investments in space capabilities, particularly in SSP will make significant new energy—clean energy—available globally. In many instances in the emerging economies, the large-scale use of renewable energy (e.g. solar arrays) is not viable due to local seasonal variations in the availability of insolation. For example, in many places it may rain for weeks at a time: how can solar be relied upon in such locations? With SSP to provide power when the Sun is not shining, ground renewables can be pursued far more aggressively without negatively impacting the local economy. Moreover, SSP will contribute directly to both United Nations Sustainable Development Goal

(SDG) number 7, "Ensure Access to Affordable, Reliable, Sustainable and Modern Energy for All", and to SDG number 13, "Take Urgent Action to Combat Climate Change and its Impacts". Providing power to schools, hospitals, businesses and homes, SSP will also support the growth of poorer countries' wealth and will help to reduce social and economic disparities across and within countries.

Commu
and
on the M

nicating

Moving

oon

**NASA's Artemis program, which includes
everything from landing astronauts on the lunar
surface to eventually building an outpost there
akin to the research stations in Antarctica,
will for obvious reasons be utterly dependent
on space-based technologies.**

One advantage Artemis has that Apollo did not is that many of the necessary technologies for sustained settlement have already been developed for use on Earth. Communications constellations, for example, are planet-agnostic. Starlink in orbit around the Moon or Mars would work equally well.[40] (This is one reason why SpaceX— if, indeed, the company is serious about Mars settlement—might have invested in Starlink in the first place.) Such a system would necessarily need to communicate with Earth, both with mission control and as part of the Internet.

In addition, ground navigation will be a vital element of any exploration of any celestial body. In Antarctica, expeditioners easily lose their bearings and sometimes die because natural features and artificial structures are few, and the human mind thus has a terrible time assessing distance and direction in such landscapes. This problem is mitigated with global positioning systems. No such satellite constellation, at present, exists around the Moon. In lunar sunshine, this is rendered less of a problem because of the indelibility of footprints and tire tracks in regolith. On a longer timeline, however, navigation from one base to another is an issue that will inevitably need to be addressed, as is the navigation from a base to any feature worthy of study, including the permanently shadowed regions of the Moon where there is no sunlight.

**Perhaps even more so than on Earth, where
myriad modes of communication and navigation
exist, a permanent human presence on the lunar
surface requires a complete installation of robust
and durable space-based infrastructure.**

Artemis taken to its natural conclusion will yield the equivalent of a ground-based International Space Station, with mine and refinery. Astronauts and robots will need to build everything from fully-featured habitats to greenhouses (and the food grown within). As the Moon is developed, so too will a local economy there. Already, terrestrial organizations such as Telespazio and the European Space Agency are, by way of feasibility studies and preliminary design efforts, looking ahead to achieving this multi-planetary infrastructure.[41]

The Artemis program is a multi-pronged effort with "buy-in" opportunities for every signatory of the Artemis Accords. Rockets are key, yes, and capsules and habitats, but things like Lunar Gateway[42]—a space station that would orbit the Moon—and Earth-and Moon-based systems critical to supporting the enterprise, are also vital. Even such elements as a landing area for rockets and transportation infrastructure from landing site to Moon base must be constructed. Again, the nearest analogs modern humans have of such a logistical enterprise in an austere region are the research stations in Antarctica, a challenging continent not yet subdued.

Establishing any sort of small base on the lunar surface, regardless of scale, would by default be one of the great engineering achievements of humankind.

And the engineering might be a lot different than traditional such projects on Earth. For example, the Moon has lower gravity, which must be taken into account, which means new manufacturing methods might be necessary. In terms of science, Moon-based biology and medicine would need to be rethought. Because the lunar surface possesses many of the elements present on Earth, in-situ resource utilization—that is, using the materials of the Moon to sustain a human presence there—would be a critical enabler.[43] There is water ice on the Moon, and there is valuable material in the lunar regolith.[44]

In such a new and untapped environment are thus enormous opportunities for businesses. There is money to be made in space tourism, for example: an "easy" short term goal. But given that

Elon Musk's goal with SpaceX is the colonization of Mars, there might indeed be a powerful, long-term deep space economy in the making.[45] With respect to the rare Earth elements that might be extracted from the lunar surface, it will take time even to determine what such business cases might be. But the basics are sound for the fundamentals: air to breathe and fuel for flying. The simple extraction—and here, "simple" is a relative term, for it is quite a challenge indeed—of hydrogen and oxygen would yield water, air, and rocket fuel useful for both a Moon base and for pushing deeper into the solar system.

Humans are well positioned for transcending our bonds to Earth. In a sense, we have been preparing for that since the dawn of the species. But just as we now look at terrestrial exploration and expansion through a lens of sustainability, so too must our expansion into the darkest ocean.

A Conversation with

Paolo Gaudenzi

Since 2000, PAOLO GAUDENZI has been a full professor of Aerospace Structures and Construction, and head of the Mechanical and Engineering Department at La Sapienza University (Rome). Previously, he was founder and director for twenty years of the Professional Master in Satellite Systems and Products, and founder and CEO of Smart Structures Solutions, a spin-off of La Sapienza University for four years. From 1991 to 1992, he has been visiting scientist at the Massachusetts Institute of Technology (MIT).

Could you share with us when and how your interest in space exploration was born?

Apart from the steps of my academic career in the aerospace research, a rich and complex environment, I was a passionate fan of space exploration since my early years, namely during the 1960s and 1970s when the Apollo program deployed the mankind's largest effort ever in terms of human exploration of space, reaching the Moon.

Can you explain how satellite telecommunication and navigation work?

Since decades, telecommunication (TLC) satellites positioned in the geostationary orbit have been offering radio signals receiving and transmitting stations, for example for TV broadcasting, telephone and Internet purposes, assuring a very large coverage of the Earth surface and fixed connection points in the sky for users. Navigation (NAV) space constellations—as GPS or Galileo—provide instead accurate positioning on Earth for moving objects, a service available on different terminals, including smartphones.

Basically, a single satellite (in case of TLC services) or a constellation of satellites (in case of NAV services) are used as a point of connection between different points on the ground or different points in space to receive and transmit signals associated with electromagnetic waves. In case of TV broadcasting, for example, signals coming from a ground station containing all the information concerning one TV channel are received by the satellite and then transmitted back to Earth, reaching very large portions of the ground and enabling users to access that TV channel. For NAV services, instead, signals coming from the ground need to reach at least three satellites before the NAV system can measure very accurately the distance between those satellites and the user's device thanks to the emission of signals and the evaluation of the path they perform.

What benefits can we derive from satellite telecommunication and navigation services?

One of the most important benefits that humans can currently get from these satellites is linked to the widespread coverage that they can provide. Covering Earth areas that cannot be reached by terrestrial means, satellite TLC and NAV systems indeed provide the solution for the digital divide, that is the gap between those who have access to information and communications systems and those who do not. Given the increasingly pivotal role of connectivity for human activities, this issue is more pressing than ever to ensure inclusive social and economic development.

Do other applications of satellite telecommunication and navigation exist besides terrestrial uses?

The need for communicating positioning information is not limited to Earth only. Space missions and activities on other celestial bodies, including the Moon, require similar services. Communication between satellites and ground bases is essential for space systems to be operative, as well as for all possible activities carried out in space, which need a communicating system backbone for the most different utilizations. Notably, this holds true for low Earth orbit activities that are conceived today not only in direct relation to Earth, but also for executing in orbit services (IoS) such as accessing to, repairing, refurbishing, removing or displacing spacecraft.

You mentioned the Moon. Can you expand on the possible development and utilization of satellite telecommunication and navigation services for the advancement of lunar activities?

The activities planned as part of the forthcoming Artemis missions envisage a permanent human presence on the Moon. This will imply to arrange for a variety of operations essential to make it possible, such as habitats' construction, energy gathering, food production, environmental conditioning of human habitats and many more. This means that one could well expect a Lunar space economy to be developed as soon as the pace for permanent activities on the Lunar surface increases. At the same time, the creation of a Moon outpost and all the complicated operations needed for the purpose, including the deployment of the Lunar Gateway, have caused a huge need for TLC and NAV services.

Do you know any ongoing project committed to the development of satellite telecommunication and navigation for lunar activities?

In Europe, there are several under development projects concerning the design of TLC and NAV services, both at the European Space Agency (ESA) and at the National level. Italy, for example, is assuming a key role in the context of Artemis missions not only thanks to Thales Alenia Space Italy, which is manufacturing components for the Lunar Gateway and for the infrastructure to be deployed on the Moon's surface, but also thanks to Telespazio, which is running feasibility studies and preliminary design efforts on TLC services. Notably, both companies are also part of the Moonlight Initiative, the ESA's project aimed at developing the TLC and NAV services.

What technology solutions will be required to achieve the Artemis program objectives? Are they already available?

Actually, every space technology will be interested in the Artemis missions, which will require newly developed space transportation systems and space segments for the Lunar Gateway and all the Earth-based and Moon-based stations. For many components of the complex systems that will operate the lunar missions the development effort will be huge in terms of system capabilities and operations. In some cases, as for new launchers, most of the required technologies are already available and not much different from previously adopted systems, although they will be used for the implementation of a new and more powerful transportation system. In other cases, such as for construction and assembly activities in space and on the Moon's ground, new advanced technologies, such as additive manufacturing, will be deployed.

What will be the addressable "uses" of the Moon?

We should focus on establishing a permanent presence of humans on the lunar surface. In fact, this goal stands at the basis of any other objective, assuming that missions not based on a continued presence of human beings will be extremely difficult at least in the first phases of the new wave of activities on the Moon. It is also difficult to define what the term "use" does mean. However, humanity might benefit from a wide range of activities that can be performed on the Moon. Firstly, the presence itself of a Lunar basis will generate opportunities in terms of observation of outer space but also (why not?) of the Earth and the surrounding environment, as well as possibly of all the nearby flying objects. Secondly, the reduced level of gravity might enable new scientific experiments and discoveries, in the field of biology and medicine for example, as well as material production and components manufacturing activities. Thirdly, the utilization of natural resources available on the Moon, like underground water or other resources embedded in the lunar soil, the so-called regolith, will require a lot of effort and consequently a lot of new activities fostering new business opportunities. Nonetheless, real business cases are still too far in time to predict the possible evolution of lunar activities in terms of commercial use. For sure, the Moon's resources will be critical for developing space activities such as the generation of hydrogen and oxygen for space transportation systems conceived for travels from Earth to the Moon and back, or from the Moon towards Mars and deeper space destinations.

How do you think that LCNS will be regulated?

I think that the Artemis mission, being the largest and most important mission of the near future, will drive and set the standard for the performances and operations of the next LCNS missions and somehow dictate the regulations for these services in their first configurations. However, developing fair and conducive regulations will be challenging. Just think that the International Space Station, although developed in a time when people's determination to cooperate in space was evident —as common sense would suggest at any time-, has required a significant regulatory effort. Nowadays, the cooperation between space nations appears in some cases weak, to be on the positive side. Then, we will probably see different actors deploy parallel LCNS to allow independence of access and operations.

What are the main benefits for Earth derived from the investments on LCNS?

Other than mentioning the wide range of services that Earth activities might derive from spin-offs of technologies created for the Moon, as well as the economic development resulting from a lunar market, I would like to point out an additional perspective on this. Humanity has always been fascinated by the exploration of unknown grounds, and space does not make an exception. If space exploration activities emerge as the natural expansion of activities on Earth, there should not be a strict separation between space, or Lunar, activities and Earth ones, and therefore between related benefits.

What Sustainable Development Goals (SDGs) can benefit the most from LCNS?

The Sustainable Development Goals should and will influence all industrial activities. On one side, one should consider that space activities will follow the trend, and, on the other side, it is reasonable to expect that space activities, starting with Earth observation, will greatly contribute to the sustainable development of planet Earth and of all mankind. In this framework, all the technologies that will be developed or advanced to serve on the Moon will provide relevant spin-offs that will possibly enable new terrestrial applications. For example, lunar exploration programs and related LCNS might lead to valuable outcomes in all the disciplines related to space biology and space medicine, which have paramount importance for deploying a permanent lunar outpost. Through ad hoc adjustments, the communication support network that will be offered by LCNS might play a critical role on Earth as well, for telemedicine, personalized medicine and all the related treatments.

Space
Circu
Technol

The most expensive, and often riskiest, part of space exploration and infrastructure development is the rocket launch.[46]

Pushing people and tonnage into space using persistent, controlled explosions is an almost comically challenging way to start any journey—and the troubles don't stop there. Once off-world (and particularly in the case of multiplanetary activities such as building a permanent human presence on the Moon or Mars), celestial dynamics must also be factored into the mission.

For example, spacecraft can only travel to Mars every twenty-six months or so, when the Red Planet and Earth are properly aligned. Before humans could reasonably claim to have built a colony on Mars—or short of that, claim even to have established a beachhead—multiple launches carrying astonishing amounts of equipment, food, and life support essentials, will have first been required to fly with, or ahead of, settlers.

One way of ameliorating the supply problem is to reduce, reuse, and recycle everything in space, and when possible, to use the existing resources on another world as raw material to build the habitats themselves. For such projects, there is little margin for error. Engineering and exploration triumphs aside, a Mars colony would inescapably be the single greatest achievement in sustainability ever executed.

To wit, the farther you are from Earth, the greater the need to reuse and recycle every atom brought on the journey, and to find ways of manufacturing new things from the celestial target in question. Launch prices are measured per kilogram, which means NASA cannot afford to fly bricks and steel girders from planet Earth to planet Mars.

Any building materials necessary for any infrastructure of serious scale will need to be extracted from the ground—whatever ground on whatever world that might be. Likewise the water, air, and fuel required to keep people alive and get them back to Earth. There is no long term timeline of exploration and settlement that does not, ultimately, require a colony to be fully independent of the Earth. Life support systems will thus need to be a "closed loop".

The International Space Station has proven that such systems can be established on a small scale and work for extended periods. Urine, for example, is recycled as drinking water for astronauts on board. Astronauts, however, are highly trained and disciplined, understanding fully how vital it is to waste not a single molecule if they can help it. And, practically speaking, their every move is monitored by NASA. As populations grow off-world, such training might not be so expert—but discipline will still be the price of entry for anyone who hopes to leave Earth for long durations.

The principles of sustainability will necessarily be fully integrated into the human psyche if humans hope to reach beyond Earth.

From raw materials to human behavior, supply lines have always limited human progress, whether building a colony in Jamestown or building a colony on Mars. Moving cargo off Earth is particularly hard, which means any technologies that improve recycling processes or capabilities are prized by the nascent space industry. And like much research that goes into space exploration, money spent to get people off Earth pays dividends to those who stay behind. The sensors, automation, and artificial intelligence necessary to build a "closed" chemical, biological, or material system can be used for similar purposes here as well. Beyond that, it can prompt a shift in how we view life on Earth.

If and when humankind engages in such a zero-waste endeavor as off-world settlement, the very definition of the word "waste" would need to be reconsidered. If, indeed, every molecule of waste is viewed not as something which must be discarded, but rather, as something that must be purposed to something productive or constructive—and if the concept is *demonstrated* as opposed to being merely notional—it would go a long way toward teaching humans on Earth how to be better stewards of our home planet.

And the ability to build or grow things using minimal resources, all recycled, is only the starting point.

Any discussion of space exploration inevitably gets mired in the technical, as though humans were mere automatons, or sparkless matter in a complex system—needy cogs in a large machine.

In fact, the very thing that drives us to explore the stars must itself be nurtured: our spirit, our senses of longing and joy and hope and pleasure. Food and water are vital to keep humans alive, but the act of breaking bread is itself a sort of nourishment.

This was perhaps best shown through the European Union-sponsored research project Eden ISS. It was intended to be a sort of terrestrial test of plant cultivation systems with applications for eventual space exploration, and involved a greenhouse built in Antarctica that was to grow food without soil or natural light.[47] The idea was to find which plants fared best in harsh agricultural conditions. Not only did the project achieve its botanical goals, but it achieved psychological ones as well: one of the scientists on the project learned that the greenhouse had a profound, positive effect on the disposition of the Antarctic-based researchers. It is one thing to eat something freeze-dried and shrink-wrapped. It is quite another to pluck it from a vine and put it in your mouth.

A Conversation with

Waltraut Hoheneder

—

Barbara Imhof

—

René Waclavicek

WALTRAUT HOHENEDER, BARBARA IMHOF, and RENÉ WA-
CLAVICEK are co-founders and the core team of LIQUIFIER
Sistems Group (LSG), a Viennese Space Architecture company that engages in cutting-edge research, design, engineering, innovation, prototype development, and testing
for space and space-related habitation, transportation, and
exploration technologies. The full LIQUIFER team is trans-di-
sciplinary and includes experts from such diverse fields as
architecture, industrial design, engineering, robotics, biomi-
metics, ergonomics, material sciences, and such manufacturing technologies such as 3D-printing.

Could you share with us when and how your sensitivity to space stations was born?

Being interested in integrating future developments into her work as an architect, Barbara studied at the International Space University in Strasbourg (France) after graduating from the University of Applied Arts in Vienna. Barbara and Waltraut have known each other since her studies at the University and, sharing a passion about space architecture, they finally became self-employed with a company specializing in that field. Some years later, René met Barbara during his studies at the TU Vienna, where she taught at Helmut Richter's Hochbau 2 Institute and offered space architecture design programmes. Therefore, we might say that our sensitivity to space habitats has a long history and traces back to our personal interests, other than to our company's core purpose.

What do we mean by "space stations"?

A space station is any kind of dwelling for humans beyond the Earth's horizon, be it in the form of an orbital platform like the International Space Station, in Earth orbit, the Lunar Gateway, an outpost on the surface of a celestial body or any kind of habitable vehicle such as a pressurized rover for planetary excursions or a spaceship for traveling to other planets like Mars.

Where are currently existing space stations positioned and where will future ones be situated?

Currently two space stations are operating, both orbiting Earth at an altitude of about 400 kilometers: the International Space Station and the Tiangong space station. The former was launched in November 2000, resulting from a joint venture of space agencies from the United States, Europe, Japan, Russia and Canada. The latter in April 2021, when China took the first step to have its own permanent modular space station. Currently, the International Space Station partners are developing the Lunar Gateway, a structure that will orbit the Moon serving as a support structure for future crewed lunar surface missions, planning to put it into service in the next few years.

Are space stations built on Earth?

Space habitats are usually modular structures prefabricated on ground and assembled on site. Often, they are launched as fully integrated units. In some cases, the outfitting takes place after the empty vessel has been installed, depending on module size, weight and launch capabilities.

What are the fundamental technologies involved in space stations?

Rocketry and shields protecting humans from deadly radiations outside the Earth magnetic field are key technologies for space travel. At the same time, with increasing destination distance, the effortful transportation from Earth forces us to focus on any technology capable of sustaining life independently from terrestrial supplies. In Situ Resource Utilisation (ISRU), namely the exploitation of all the resources that are available on site, is thus another crucial element. On-site resources might include energy and materials such as sunlight, regolith (the Moon or Mars soil that might be used, for example, to create human shelter), as well as water gasses and other elements (needed for producing propellant or life support substances). For the same reason, closed loop life-support systems, namely self-supporting controlled closed ecological systems, will be essential to establish and maintain habitable environments.

Space activities push technological advances forward. Could you mention some other technologies whose development is being boosted by the evolution of space habitats' circular systems?

Given the need for efficient resource management and avoidance of waste generation, any technology aimed at recycling or upcycling existing objects or generating new ones is highly valued and fostered in the space industry. Additive manufacturing tools, such as 3D printing, are surely a valuable example as they allow optimizing resource use, carrying out rapid prototyping and production of spare parts, and creating mono-material products suited to continuous changes of physical properties to serve different functions across time. Notably, space exploration plans are putting high pressure also on the further development of information and communication technologies, sensors, automation, and artificial intelligence applications, as crucial to realize complex self-organizing systems.

What are the major challenges related to creating and maintaining habitats located outside the Earth atmosphere?

That depends very much on the chosen site. However, other than the afore-mentioned transportation constraints and the lack of the Earth magnetic field protection, there are other common challenges. For example, further technical advancement is needed before achieving stations' full circularity. More specifically, although we can seamlessly create some resources loops on Earth, replicating them in space is still complex as they require extensive machinery, whose dimensions hinder the effort to leave as much space as possible to humans and their activities. Therefore, the real challenge concerns the miniaturization of systems and their 100% reliability. Moreover, the side effects of microgravity on human bodies is another major issue when planning habitats in low Earth orbits, namely up to 2000 kilometers above Earth. Another challenge derives from increasing distances, as we are faced with safety threats simply because a mission four million kilometers away from Earth can hardly be supported or even aborted in case of emergency. Also, given the communication delay due to the limitation of signal speed, such missions require a high level of autonomy already during the construction phase.

The concept of waste is key to circularity on Earth, how does waste affect space?

Space debris include all derelict man-made objects that do not serve a useful purpose anymore, as abandoned spacecrafts, launchers or their components, as well as fragments of any of these objects. Today, an ever-growing amount of space debris damaging orbital hardware has already become a cost factor, thus it is broad consensus that future habitation structures on celestial bodies can only be sustainable following a recycling paradigm that is aiming for a closed loop where waste products are widely avoided.

Would it be possible to use space debris or other already-in-orbit materials to build and/or maintain space habitats?

Mary Douglas wrote, "Dirt is matter out of place": we believe that space waste and debris are matter out of place as well. If we look at them not only as hazards but also as unemployed materials that can be recovered and devoted to new purposes, our approach to their management can change. Today, required solutions to space debris crowding, the need to lower the high costs of materials transportation from Earth to its orbits and beyond, as well as the new plans about establishing human settlements on the Moon and other celestial bodies, all converge on the possibility to produce with already-in-orbit matter. In fact, this has already happened when the concept of reusing logistic hardware has been taken as a key principle of the US Space Shuttle program, and new solutions are continuously emerging, such as, for example, SpaceX's recently developed reusable rockets. In our LAVA HIVE project, in 2015, we saw in re-entry shields and other landing hardware, which are normally abandoned immediately after they have served their scope, valuable materials to build a hardtop where life support systems can be positioned.

What is your vision of future space habitats?

Future space habitats in planetary surfaces will have to be built using local materials. Today technologies for producing building elements from local sand/regolith are already being advanced. We will also use local topological features such as caves, lava tubes or canyons to shelter our habitats from harmful radiation and, in the case of the Moon, from micrometeoroids.
We can imagine the first bases on the Moon to be organized like bases in Antarctica. On the Earth pole, they have living quarters for two types of crews: crews that are staying there for shorter duration, between one to three months, and crews that are overwintering there for nine to twelve months. These inhabitants do research and bring their research equipment or install it to conduct their work. We can imagine something similar on the Moon.

What have you learnt about Earth working on space habitats?

Working to overcome space habitats limits makes you very conscious of what is strictly necessary for human beings to survive, both in the short and in the long term, and how to make the most out of it, avoiding any potential waste. Coming close to this consciousness can be an eye-opening experience on more responsible approaches to resources use on Earth.

Reading about space, the term "spaceship Earth" can occur quite frequently. What does it mean?

The phrase was coined by Richard Buckminster Fuller to describe our planet. He was an architect, but also a philosopher, an inventor, a designer, a professor, and even a TV presenter, and observing our world he felt that all human beings were passengers on a giant spaceship, Earth, and, like the crew of a large ship, people had to work together in order to keep the planet functioning properly. Buckminster concerned himself with humanity at large, and with this formula he took a big-picture view of the world and its problems.

You have worked on the concept of "City as a spaceship", what do you mean by it? What is your vision of future Earth habitats?

We believe that current space exploration missions can offer many insights on how to develop more sustainable terrestrial urban areas. As we are facing further population growth and increasing urbanization, new approaches to cities planning and management should be embraced. Evoking Buckminster Fuller's words, we would like to propose a new urban philosophy, which envisions cities like modular closed loop ecosystems, consisting of many littler spaceships both self-sufficient and capable of sharing excesses and exploiting economies of scales and efficiencies. Exploiting technologically advanced infrastructure, we hope to see megalopolis like Bangkok, Buenos Aires, London, Mexico City, Moskow, Mumbai, New York City, São Paulo, or Tokyo transform and opt for compact, multifunctional and public-private spaces, renewed resource, waste and health management, alternative energy harvesting, and inclusion of nature into built-up environments. We love to think about cities as spaceships primarily to tackle the needs of an increasing and resource-poorer population and to ensure better living standards for all.

Mining
ogies

One of the oldest tropes of science fiction sees space travel as routine and unpleasant, brimming with the sorts of jobs men and women get when they can't do anything else. And of these jobs, few are more prevalent in fiction than "space miner" on the asteroid belt. Sometimes robots help out, as Isaac Asimov envisioned, and sometimes the race beyond Mars is a kind of echo of the California Gold Rush in the mid-nineteenth century, as Robert Heinlein imagined.[48] But the idea of entire planetary objects teeming with gold and diamonds is inherently understandable even to people who have little interest or knowledge of space. After all, the asteroid belt is just a planet that never quite came together, and should have all the precious metals and minerals hoarded on Earth. One gold mine on the Moon could pay for a thousand Apollo programs—right? Maybe, but not any time soon.

There are two chief cases for space mining: harvesting resources on other worlds for use on Earth, and doing the same thing, but for use in space. Perhaps unexpectedly, however, the case for mining celestial objects to return materials back to Earth does not make a lot of sense, economically—at least, not yet.

Simply put, it is very, very expensive to extract resources from other worlds, and, relatively speaking, pretty easy and cost-effective to do the same thing on Earth. Yes, rare Earth elements are ever in demand, and yes, geopolitical strife makes getting legal access to them a challenge, but the sheer size of the Earth is inconceivably large to the human mind, and though metals and minerals are hard to get to, they are still in great supply, and we are getting better at reaching deeper into the Earth to extract them.

While mining celestial objects for resources that we will use on Earth will happen in the long term, it's not something one should look forward to happening any time soon. The cost of extraction,

whether financial or economic, would have
to exceed the cost of space excavation
and return to Earth.

That doesn't mean, however, that we won't one day—and perhaps one day soon—be mining the Moon for use on the Moon. As early as the 1960s, NASA began studying how Moon mining might work.

The decades since have seen satellites scanning the lunar surface for potential elements, including those necessary to make rocket fuel. And the next few years will see the Artemis program land technology demonstrators to dig some of it up.[49] Such uses make perfect sense. Water, to name one human necessity, is incredibly heavy, and every kilogram of mass launched from Earth costs thousands of dollars.

If there are oxygen and hydrogen molecules on the Moon, NASA, ESA, and other space agencies involved in the Artemis program would find a better return on investment by building machines that over time can produce metaphorical rivers from lunar regolith. By saving money otherwise spent on launching from Earth resources already found on the Moon, space agencies could invest in other areas such as transportation, new bases and colonies, and more robust supply lines.

As a practical matter, mining asteroids, the Moon, or Mars isn't that different from mining the Earth. Prospectors find the desired resource. Engineers determine the feasibility of extracting it. Miners or mining equipment is put to the task, and it is harvested, processed, and utilized. The main difference is one of environment.

It is one of the few industries that humans have spent millennia mastering, and which apply one-to-one on other bodies. The hardware necessary is essentially the same on the Earth or Mars, but the environmental concerns—in other words, gravity and temperature—are different, and that is the area to which such equipment must be adapted.

Before any of this can happen, work remains to be done on the legal framework for such mining activities. Presently, no such framework exists. However, multiple groups, including the Hague Space Resources Governing Working Group and the United Nations Committee on the Peaceful Uses of Outer Space have begun hammering out what such regulations might look like. Moreover, individual countries have begun the laborious legal process of working through who might own what if it were hewn from a celestial object. The Artemis Accords amount to an early, successful international effort to mine the Moon and other bodies in an orderly manner.

A Conversation with

Angel Abbud-Madrid

ANGEL ABBUD-MADRID is Director of the Center for Space Resources and the Space Resources graduate program at the Colorado School of Mines in Golden, Colorado, USA. His work focuses on the study of all aspects of space resources, including their identification, collection, extraction, and utilization.

Could you share with us when and how your interest in space resources was born?

My interest in space resources was born from the subject that I worked on during my doctoral research work. This was related to the combustion of metals on the Moon and Mars to be used as rocket propellants. Afterwards, I joined the Colorado School of Mines in the late 1990s, where I became part of a new research group interested in conducting studies of resources beyond our planet.

What is space mining and what are promising celestial bodies to mine?

Space mining can be defined as the extraction of any elements beyond Earth—solid, liquid, or gas—for utilitarian purposes. These resources include water, oxygen, gasses, volatiles, minerals, metals, and regolith. In the near term the most promising celestial bodies for space mining are the Moon, Near-Earth asteroids, and the planet Mars. These desirable destinations have valuable resources in situ for a variety of space applications, which are needed by humans to expand their in-space activities (such as consumables for human settlements, propellants for refueling spacecraft, and materials for construction and manufacturing) and whose extraction can be justified economically when compared to sending them from Earth.

Are space mining technologies similar to ones used on Earth?

Given that Earth is a planet in which we have hundreds of years of experience mining resources, early designs of technologies for mining space resources were very similar to terrestrial ones. However, these designs have evolved to take into account the different operational environments at the various destinations, such as low gravity, high vacuum or low atmospheric pressures, extreme temperature variations, radiation, meteorite bombardment, and severe constraints of mass, power, and volume with minimum interaction with humans. However, as the complexity and size of the mining equipment increases to enable a sustained human presence on the lunar surface, humans will play an important and more involved role to maintain and operate this equipment.

At what stage of development are these technologies currently?

Several of the early technology concepts from the 1960s are still around, but the engineering designs and prototypes have advanced significantly since then. At this point, remote sensing technology for identification of resources has been used extensively for several decades already, while equipment for surface exploration, drilling, excavation, extraction, and material handling and transport is currently being developed. These technologies are supposed to be demonstrated on the lunar surface during the second half of the 2020s, following several prospecting missions by various space agencies in the first half of this decade to locate, characterize, and quantify resources.

In the long-term, do you think space mining could contribute to alleviate potential risks of supply shortage and/or geopolitical issues arising from vulnerable dependence on key suppliers/countries?

While there is a growing need for critical materials and metals required for product manufacturing on Earth and their supply is in some cases subject to shortages and geopolitical issues, currently it is not economically or technologically feasible to bring them from space. Resources in our planet are still abundant, exploration efforts are constantly underway in many locations around the world to identify ore reserves, and new technologies are being developed to continue extracting metals and minerals at increasingly deeper levels. As a result, the cost of extracting terrestrial resources is orders of magnitude lower than the cost expected from extracting resources in space and transporting them to Earth. Thus, importing materials from space is not an activity that will happen in the near term or even in the foreseeable future. In the long term, it would require reaching very high resource extraction costs on Earth or deciding to avoid every environmental damage derived from terrestrial mining operations, in addition to already having a well-developed infrastructure in space to justify bringing resources from space to Earth.

If not to provide Earth with a renewed supply of resources, will space mining contribute to human activities in space?

Given that importing resources from space is not economically feasible in the foreseeable future, the main use of space resources will be in space. Their contribution to space exploration is mainly linked to the extremely high cost of launch operations, which are very energy intensive due to Earth's large gravity well and which burden space exploration missions' feasibility. If instead, resources such as water, oxygen, metals, and regolith for human consumables, propellant production, habitat construction, and manufacturing are obtained from space, this would significantly reduce the cost of having to launch them from Earth. This is what is usually referred to as the "living-off-the-land approach" in space.

Who are users interested in space resources?

Since resources are elements that have value because of their utilization, a demand or customer must exist for their use. In this context, early users will be space agencies and other government organizations interested in the development of infrastructure for lunar and planetary surfaces for robotic operations and human settlements, as well as for manufacturing of orbital structures. Such complex activities will require these organizations to reduce their costly dependence from terrestrial supplies to become a reality. As this infrastructure grows and becomes operational, the private sector will follow as a user of in-space resources for commercial opportunities.

What are the main countries involved in pursuing space mining and what is their goal?

At this point the main countries involved in efforts to develop space mining technology are the United States and China, followed by Luxembourg, the European Space Agency, Japan, Russia, India, South Korea, the United Arab Emirates, and Australia. The large impact that space resources can play to enhance space exploration efforts and increase future economic activity beyond our planet are currently driving a new space race. However, this time the objective is not just reaching a destination and proving the technological superiority of competing countries like in the 1960s. This time the intention is heading to space, the Moon, and other planetary bodies and using local resources to enable sustained presence there, while also reaping their benefits for humans on Earth.

What are the key challenges and next steps to enable space mining?

As on Earth, the extraction of space resources will not happen until there is enough geological and technological information to positively identify and quantify recoverable resources, but also until there is economic and legal certainty that resources can be considered proven reserves. Thus, the main challenge at this point is conducting enough exploration, from orbit and on the surface, to characterize the location, quantity, distribution, concentration, and accessibility of resources before space agencies and companies invest on developing equipment to extract them. Given the high cost and complexity of such thorough exploration efforts, space agencies and companies will have to initially join forces and work together to gather this important information. At the same time, a sound legal framework should be pursued and established to assure that space mining activities are conducted in a fair, responsible, and sustainable way according to international treaties.

You mentioned the need for a "sound legal framework" to enable space mining. What actions have been taken so far to address the legal issues surrounding space resources?

A formal legal framework for space mining is still not in place. However, there have been several efforts to move in this direction. Such is the case of the work done by The Hague International Space Resources Governing Working Group, the Global Expert Group on Sustainable Lunar Activities (GEGSLA), and the United Nations Committee on the Peaceful Uses of Outer Space (UN COPUOS). There have also been laws already passed at the national level in several countries to address the extraction of space resources by the private sector. The United States, Luxembourg, the United Arab Emirates, and Japan are some examples. Finally, countries have banded together and agreed to conduct safe, responsible, and orderly space mining operations in organizations such as the Artemis Accords with currently more than 20 countries as members and the International Lunar Base with China and Russia as its main partners.

The Artemis Accords gather States around a common agreement to operate according to the principles of the *Outer Space Treaty*, a set of guidelines issued by the United Nations in 1967 and representing the main international space law reference point on space exploration and exploitation. The latter prohibits national appropriations of outer space. How are property rights addressed in the field of space mining?

Property rights in terrestrial mining developed after a legal framework was in place. Given that this framework is currently lacking in space, because of conflicting views on how to interpret international treaties on the use of outer space, property rights remain *terra incognita*. However, government agencies prompted by the private sector are increasingly realizing that property rights will have to be established for an effective, fair, responsible, and sustainable extraction of resources to enable both future scientific exploration and commercial activities in space.

It is a common opinion that investing public funding in space should have a low relevance in the agenda of governments as multiple economic, social, and environmental issues on Earth need to be solved, first. What is your opinion about this statement?

While space activities may be perceived by some people to benefit just a few countries and companies, and in the future, just a few human beings who may live in lunar and planetary bases, the vital importance of space on the daily lives of human beings on Earth is undeniable. In the future, economic activity deriving from space activities will continue to benefit the billions of human beings who inhabit this planet and who will never leave our cocoon. So, while most public investments should go to the multiple economic, social, and environmental issues that we have on Earth, it is clear that investments in space technologies, albeit small in comparison to terrestrial ones, will continue providing important and impactful benefits to our well-being on Earth. As for space resources, just like resources on Earth which are the engine that generates all products and services for our technological society, they will become the enabling element behind future space exploration, scientific discoveries, technological innovation, and commercial activities that will help incorporate space into our ecosystem and global economy.

Space
Sust

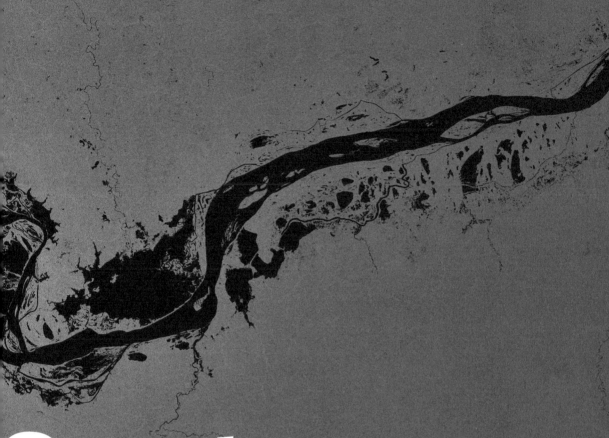

Sector
...ainability

A question worth asking is what must come first
when building sustainability into space exploration.
The answer: Everything.

Space based power systems will be of no benefit to Earth if an
orbital solar collector is obliterated by space debris. There is little
purpose in going through the expense of building a lunar colony
if circular systems are not mastered. Mining the asteroid belt for
lithium to use on Earth is ultimately meaningless if we are not
monitoring the Earth from orbit and working on the ground to
clean up our warming world.

Space sustainability is not particularly complicated as a concept.
In a sentence, it means preserving or improving the condition
of space (and access thereto) for ourselves and the future. It is
the sort of general notion with which no sane person would dis-
agree. Indeed, there is a famous Boy Scout rule: "Always leave the
campground cleaner than you found it". If all players in space
today, whether national agencies, militaries, or billionaire tourists, followed the space equivalent of that rule, any outstanding
issues of space sustainability would resolve themselves in a single
generation.

It is, however, an oversimplification of the problem, and the met-
aphor is imperfect. When Boy Scouts go to Yosemite, they do not
look around and start planning the city they would like to build
there.

Aside from a couple space stations and a few constellations of satellites, space infrastructure as yet does not exist. For the first time
in human history, however, technical capability has aligned with
national and corporate interests and will. The last time such an
alignment occurred, and on such a stage of unbridled potential for
civilizational transformation, James I was king of England.

We have the tools and the money to build meaningful infrastructure in low-Earth orbit, cis-lunar space, on the lunar surface, and
on Mars. But if there is anything the settlement of the New World
taught us—indeed, the settlement anywhere at any time since humans began walking upright—it is that building infrastructure is
messy. If not handled carefully, it can be the antithesis of sustainable.
Adding to the problem is that since 1957, space programs of the

world have never really considered, or at least, put little value in, sustainability. Which means the concept is less intrinsic to the endeavor than it is "bolted on", and as such, will require a shift in thinking to place it among the first principles of engineering during spacecraft development.

Space debris is the immediate problem that everyone thinks of when sustainability is discussed.

Physical access to space, and the preservation of a metaphorical global ocean at three hundred fifty miles up, is vital for anyone at all to participate in space exploration, whether a company with ambitions of colonizing Mars or a nation-state with a single satellite just launched to orbit. It is also easy to understand. Humans have littered orbit, both incidentally and intentionally, and that is, fundamentally, poor stewardship of the Earth. We don't like it in our parks. We should not like it in our skies.

Once you travel beyond low-Earth orbit, however, sustainability remains a stubborn issue. Permanent settlement of the Moon may create waste of every kind, and a pristine landform may be marred over the course of its development. Likewise Mars. So how might anyone build such challenging infrastructure in a responsible way—particularly when humans have shown little inclination, historically, to build comparable infrastructure in the relatively simple environs of Earth? Philosophically, it can get more complicated than that: What is infrastructure? What is responsible? Should nations early in the development of their space program adhere to principles that chiefly concern countries like the United States, which already has *de facto* dominance of the domain?

The most likely path to success in terms of integrating sustainability principles into all areas of space sector development is simply for nations to take it upon themselves to do it the "right way", and through international partnerships, bring other nations into a "clean space" mindset.

It is unlikely that the United Nations can develop overarching guidelines with teeth that everyone will sign on to. Waiting for that to happen is folly. While the climate debate is mired in disagreements on models and assumptions, simple, observable laws of physics are driving the need to get serious about sustainable space.

Consider that even given the paucity of spacecraft in orbit around the Moon, in 2021, the Indian Space Research Organisation's Chandrayaan-2 orbiter nearly crashed into NASA's Lunar Reconnaissance Orbiter. If the Artemis program continues apace, and as the United States and other nations begin exploring the Moon *en bloc*, the near-miss might be less of an anomaly as it is a harbinger of things to come. The same will be true for Mars (though NASA is already working to address that issue with the Multi-Mission Automated Deep-Space Conjunction Assessment Process).[51]

And while bad actors on the world's orbital stage are making things much, much more difficult from a space debris standpoint, there is at least a hint of a silver lining: the Russian obliteration of KOSMOS 1408 focused nations and industry on the problem of space debris, and its threat to the business case of space development. It is one problem of many, but it is the immediate problem, and people are watching. Irresponsibility in low-Earth orbit would not garner levels of disapproval similar to, say, detonating a nuclear weapon on a battlefield, but it would certainly invite denunciations akin to someone dumping toxic waste in Antarctica. Corporations know that if they don't take the lead on this, the government will—and the government is not always known for its light touch when it comes to regulation. Moreover, every nation in the world now knows that an act similar to what Russia did would be a major international incident harming their global standing.

All this is good news. (Well, aside from the satellite debris.) It means actors in the space arena accept the space sustainability issue, and are prepared to use teeth to punish those who do wrong.

A Conversation with

Kevin O'Connell

KEVIN O'CONNELL has served in many US Government roles, working for the Department of Commerce, where he has been Director of the Office of Space Commerce, the Department of Defense, the Department of State, the National Security Council, the Office of the Vice President, and the Office of the Director of Central Intelligence. He was the first Director of the RAND's Intelligence Policy Center, where he also worked as researcher and professor, and served as an adjunct professor at Georgetown University' School of Foreign Service. He also founded two tech- and space-related consulting firms.

How do you define space sustainability?

Space Sustainability is the responsibility that we have to preserve the space environment, so that we can do as many interesting and important things in space in the future as we have done to date.
In other words, it is really about preserving future options in space, to do as many exciting things as we have already done in these first six decades. Space is as much about economic growth, talent development, and innovation here on Earth as it is about exploration of the universe.

Your definition recalls the United Nations' notion of sustainability on Earth. Do they relate with each other?

I think space sustainability is an extension of the broad sustainability concept as we think about it on Earth, for sure. While these two dimensions, the terrestrial and the space one, should certainly relate with each other, people's attention to Earth-related sustainability has increased, but space sustainability has not been addressed in parallel.
The additional challenge we have in space, thus, is that we are really late: we have not been as deliberate as we have been down here; and frankly, we have not been so great on Earth either. Such a delay also derives from the general public's still limited awareness of space-related issues, in general and, more specifically, about space sustainability. While we still expect to receive considerable benefits from the use of space here on Earth, we do not speak about them enough to make people better understand the value of space and why we should dedicate resources to explore and use it sustainably. The general public's awareness of space sustainability would improve if there were a clearer understanding of space benefits here on Earth.

What are the major challenges related to the space sector sustainability today?

I think the very first one is space debris. It is quite incredible, for example, to think that we still have pieces of the early Sputnik satellite in space. The second is that we do not have agreement on which policy paths to choose between either make space sustainable by cleaning it periodically, in the one instance, or keep it sustainable by improving awareness in space and underlining what rules should operators use in space as they're navigating.

My view is that the private sector will probably create faster, practical paths than those the current geopolitical circumstances will allow between governments.

Do you think that the space sector faces any challenges within the Earth atmosphere?

Maybe not just challenges as much as opportunities. The easiest and cheapest solution for space sustainability is not creating new troubles once you launch.

Pursuing this specific goal generates a wide range of opportunities: now we see tremendous research in alternative fuels, propulsion mechanisms, lightweight mesh to protect satellites during their in-orbit missions, and systems for faster deorbiting, just to mention a few. In each of these fields, there are a lot of opportunities to leverage. A near-term challenge, instead, is that we are going to see a growing crowding problem transitioning from ground to air to space. As we launch rockets, there is going to be an increasing amount of air traffic activity, due not only to spacecraft, but also to the so-called HAPs (High Altitude Platforms). Since we are going to see a bigger and bigger number of legitimate uses for all of those, it is just going to get more crowded vertically.

The contemporary discussion on sustainability on Earth includes not only environmental aspects but also social ones. Do you see any such challenges in space?

I think that the social sustainability theme in relation to space really begins with creating a more thorough understanding of space matters, including how much space affects all of our daily lives. When thinking about space, people present two major biases.

First, we tend to frame those who go to space only as billionaires or astronauts. On the contrary, space is really for everybody, and everybody can participate, of course from different walks of life and exploiting different disciplines. Secondly, people are mostly unaware of the benefits that they derive from space.

Some time ago, during a lecture I gave in Tokyo, somebody told me I would need to keep it simple because not everybody understands space. To immediately disprove this belief, I started by asking whether anybody had checked the weather that morning, or if anybody had used an ATM machine, had made a Zoom or Teams call, etcetera, demonstrating how space was already deeply present in their daily lives. As a final point, we do need to give people a sense of our struggle on this topic. For example, with Dr. Moriba Jah, we have been pushing an idea for a World Space Sustainability Day, to make sure people recognize that what we have in space is not a guaranteed, endless resource.

Another big theme regarding space sustainability is international cooperation. Diplomacy and cooperation seem to be disrupted on Earth due to spreading conflicts. Do you think that the future of space will be kept out of Earth tensions?

No, I do not believe that. Space has always been integral to information gathering and warfighting. The First Gulf War was the first space war, and many, many parties—including Europe, China, Russia, and India—recognized the lessons that space could be a key enabler and means of future warfighting, both for offense and for defense.
This still holds true. To note, starting from the Russian anti-satellite test in 2021, the narrative about space economy and space commerce has kind of shifted lower, and the security narrative has again jumped over the top. To think about this from another perspective, we can discuss to what extent security in space will encourage commerce, almost like what happens with the high seas: the US Navy exists to protect commercial shipping lanes; in the same way, the US Space Force exists to encourage and protect space commerce. So, it is a pretty complex topic, but by no means is the idea of security in space going away and we will have all the complications in space that Earth has.

Today there are many voluntary guidelines about how space operators should behave outside the Earth atmosphere, but they seem to be only partially decisive in tackling space sustainability issues. Do you think space needs more norms, mandatory norms especially, to be preserved?

Yes. I think the proper way to do that is to continue the work at the United Nations in what the United States, and others, have called the *"Long-Term Sustainability (LTS) Guidelines 2.0* effort"; these efforts acknowledge the importance of the existing *LTS 1.0 Guidelines* while noting that the *Guidelines* are still too general to be totally effective.
On one hand, some norms can be issued very routinely. I am thinking about norms as simple as the highway code, like for example, the rules addressing

the risk of collision between satellites. On the other hand, binding norms are much harder to define and agree upon, and ensuring compliance is even more challenging.

Just think that not even all the *LTS Guidelines 1.0* are totally agreed upon by all countries, with some only partially complying with them. Including mandatory rules implies instead defining how to penalize an actor that says "No" to them. Still, I think we need both kinds of rules.

Do you think that commercial space operators have any role in pushing forward space sustainability or should they only wait for norms to be made?

I think the business community plays a potentially huge role here. While it is often said that business operators are not going to spend a nickel on anything unnecessary, I think that the November 2021 Russian anti-satellite test triggered additional businesses' proactivity in this area. Overall, they seem more motivated to find a solution themselves—either improving technologies aimed at tracking space assets positioning or acting more responsibly when they launch new capabilities to space.

We see many joint industry papers today highlighting opportunities for improving space safety and sustainability, as well as firms actually creating businesses around these opportunities. Facing a lack of detailed rules, the basic, practical rules will largely come out of space operators who have to stay out of the way of others, for reciprocal interest and in the interest of business continuation.

When I took my job at Commerce, I realized quickly that the space community had been concerned about space debris for over twenty years. Even back then, it was not a new topic. I think one of the roles that I had was to try to talk about it more publicly. In the earliest days, people would say: "Are you crazy, O'Connell? Space is a big place. How could there be a problem of space debris there? There are only a few people up there".

On Earth, investors, looking to sustainability indicators to evaluate their investments, are playing a key role in encouraging companies to achieve increasingly higher sustainability standards. Do you think anything similar can happen in space?

Yes, absolutely. The investment community has now come a long way in recognizing this as the problem. One of the things that investors are increasingly doing is that when they look to place an investment, they are asking: "What are the practical sustainability implications of your business model? Do you have a debris mitigation and de-orbit plan?". And we are indeed starting to see a number of companies using sustainability tools.

Space

Debris

There are more than 20,000 large pieces of debris in Earth's orbit, weighing more than 8,000 metric tons, combined.[54] The problem is akin to an oil spill in the Gulf of Mexico or a ruptured pipeline across a protected habitat, though the public so far has yet to engage with the issue in a meaningful way. In part, this indifference is a function of poor communication by space and astronomy-oriented bodies describing the nature of the issue. People simply do not realize the implications of a debris-poisoned low Earth orbit, how the region got that way, and how that affects launches, communications networks, and human spaceflight endeavors. They do not, indeed, know just how important a role space plays in their lives. Moreover, the apathy to space debris is an example of crisis fatigue. In a time when everything seems to be going wrong, there is simply no bandwidth left for the public to be concerned about space garbage. The sustainability issue has a long way to go.

While solving the world's myriad other crises is outside the scope of this book, certainly improving science communication with the general public—whether by way of podcasts, books, social media outreach, or basic awareness campaigns—will be essential to making space debris-related problems a mainstream issue worthy of discussion for those other than the customary audience of astronomers and space travel aficionados.

An informed and interested public is also a public able to put pressure on players in the space sector who operate out of compliance with a sustainability framework. While it might seem like a niche issue, there are historical precedents to this, most notably, perhaps, being the hole in the aforementioned ozone layer in the 1980s. One need not have possessed a doctorate in atmospheric science—or even prior knowledge that such a science existed—to understand a straightforward problem: chemicals in consumer and industrial products risked increasing cases of cancer to people the world over. With breathtaking and inspiring swiftness, the world adopted the

Montreal Protocols, facilitated by the United Nations.[55] It is reasonable to assume a time in the future when the people who use satellite services and capabilities—this would include nearly everyone on Earth—are interested in, and able to put pressure on, satellite operators and clients to act responsibly in space, and converge on reasonable solutions to the extant problem.

Indeed, the profit motive can be a powerful inducement for satellite operators to clean up their own acts—and messes—given the stakes of malfeasance. The only way private enterprise can leverage the value of spacecraft in low Earth orbit is to keep that region of space free of debris. State pressure—even something as simple as mandating debris mitigation strategies in contracts for government services—can ensure compliance with such agreed-upon community measures.

Space sustainability as it relates to space debris is not merely a clean-up issue however. It is, quite simply, a numbers game. The more spacecraft we put into orbit, the more ancillary debris will result—and the greater the likelihood of incidental, secondary and tertiary debris events. (For example: satellite collisions caused by trajectory errors, hardware malfunctions, or space weather.) Developing a regulatory environment, meanwhile, is slow going. The United Nations is oftentimes considered the body best able to address this global issue, but has had minimal progress at best on addressing the broader sustainability guidelines it laid out in recent years, facing particular setbacks during the COVID-19 pandemic, the economic strife that followed, and the ongoing war in Ukraine.[56]

Moreover, though countries are obligated in the *Convention on Registration of Objects Launched into Outer Space* to register all such objects with the United Nations, it is obvious that many have simply ignored the agreement, or complied with data that is so ambiguous as to be meaningless, oftentimes because spacecraft have military or intelligence applications.[57] Details and metrics are vital to solving the problem of space debris: knowing what is out there is vital to knowing what must later be cleaned up.

There is, however, a silver lining to those eight thousand metric tons in orbit: engineers might be able to give them a second life. Specifically, if space programs can recover space debris and waste, that mass might be given new uses for human spaceflight.

The solution to debris, then, is less about grabbing material and deorbiting them to be vaporized in the atmosphere during reentry than it is about reorienting the tonnage already in space to productive effect.

In fiscal year 2021 US dollars, it costs about 2,600 dollars per kilogram to launch something into space with a Falcon 9 rocket.[58] (On the far end of the spectrum, NASA's Space Launch System rocket comes in at 58,000 dollars per kilogram.)[59] Repurposing only a small percentage of the 8,000,000 kilograms in Earth's orbit would still save tens if not hundreds of billions of dollars in launch expenses. Reuse and recycling already underpin the future of space exploration. Certainly, recycled air and water will constitute the backbone of any life support system for a long-duration, crewed deep space mission. Reusing space junk is the next, and perhaps most obvious, step, solving two problems at once: Providing raw material for use in orbit and on the Moon, and removing said material from low Earth orbit as lethal, useless mass. Engineers at NASA, ESA, and elsewhere recognize the potential, and are embracing it. The European Space Agency, as part of its Clean Space initiative, is studying On-orbit Manufacturing, Assembly, and Recycling (OMAR) mission architectures for "optimizing a satellite design for on-orbit manufacturing and assembly, as well as preparing it to be refurbished, reused or recycled on orbit afterwards".[60] On the human spaceflight front, the 3D-Printed Habitat Challenge, launched in 2015 by NASA and intended to find new methods of in-situ resource utilization, saw participants reuse spacecraft heat shields and hardware—normally discarded—for use in 3D-printed Martian facilities.[61] Any attempt at off-world settlement would necessarily be a zero-waste endeavor. It would certainly be a heroic demonstration of the mindset more must adopt back on Earth.

Space debris, as the most visible issue of space sustainability—and certainly the easiest to conceptualize—might yet have multiple paths to a solution. If one of those paths can help make humans multiplanetary, so much the better. If they can make life better on Earth, it would be the most optimum solution imaginable, and emblematic of why the space programs of the world matter in ways beyond the material.

A Conversation with

Moriba Jah

MORIBA JAH is a space scientist, an associate professor of Aerospace Engineering and Engineering Mechanics at the University of Texas at Austin, and co-founder and chief scientist of Privateer. Previously, he worked for the Air Force Research Laboratory, for NASA's Jet Propulsion Laboratory, where he was a spacecraft navigator, and for the US delegation to the United Nations Committee on Peaceful Uses of Outer Space. He is now fellow of many space-related organizations, a full member of the International Academy of Astronautics, and co-editor for the Acta Astronautical Journal.

Could you share with us when and how your interest in space debris was born?

My interest in the topic was born in 2006, when I left NASA and ceased to work on Mars missions. At that time, I moved to Maui, where I had the opportunity to dedicate my studies to observing the sky more thoroughly. When I looked through telescopes, I saw that there were 1,200 objects working in space and there were 26,000 pieces of garbage. That is when I became interested.

Could you tell us what are the major obstacles that are currently hindering space pollution prevention?

I would say that the hardest thing to fix is the lack of empathy. When informed about space debris, people tend to say, "That is not my problem", and this attitude probably results from a lack of empathy for the issue. The most relevant challenge is thus understanding how to convey the story about near-Earth space pollution in such a way that recruits empathy from people and pushes aside those "That is not my problem" attitudes. For this reason, creating effective and engaging science communication and social networking sites are the key to making space debris-related problems more mainstream, especially to address people that are not in the space sector and that are not aware of the risks derived from space debris.

Do you think that a more vivid risk perception from space operators could be a game changer?

It is ridiculous to just let companies decide on what they feel is risk. According to United Nations' conventions and treaties, governments are responsible for providing authorization and continuing supervision on space activities. This is mentioned in the *Outer Space Treaty*, Article VI, and in the *Convention on Liability for Damage*, which provide that launching states (defined as States that launch or procure the launch of an object in space, or from whose territory an object is launched) are responsible for all the damages that any given space objects can cause. When objects can be traced back to multiple launching states, the latter share liability, and that liability is nontransferable according to international law. When I try to talk with the companies about liability and damage, that does not go very far because there is no legal leverage. Companies simply follow the laws established by the countries that authorize them to operate in space. Therefore, governments need to impose legal actions and only public pressure can induce them to do so. This is why public awareness is the most important thing.

What do you think about the US Federal Communication Commission's new rules requiring satellite operators in low-Earth orbit (LEO) to dispose of their assets within five years after completing their missions (shifting from the currently-in-place "25 years rule")?

Right now, only a few satellite operators purposely deorbit their assets at the end of their useful life. It must be considered that given the low Earth orbits' altitude (LEO encompasses Earth-centered orbits with an altitude of 2000 kilometers or less), if a satellite dies, in five years it will naturally re-enter the atmosphere. That is like putting trash on your front porch, because eventually the wind will blow it to get rid of it. That is abandonment. So, we need to stress that uncontrolled re-entries of any type, whether it is rocket or satellites, are irresponsible! It is just abandonment, not disposal! Disposal is when you intentionally deorbit your object and similar initiatives, and that is where we want to go in terms of prevention of pollution. I am a strenuous advocate of developing a circular economy in space, which should focus on pollution prevention and on minimizing single use of satellites and rockets, avoiding abandonment by any means.

Then, what key actions should governments take to incentivize responsible actions from private actors?

If private actors continue to behave in an uncoordinated fashion and do whatever they want, their ability to capitalize on space in the long-term will be hindered. I believe that people are generally greedy. Thus, persuading satellite operators that behaving sustainably can be more profitable than abandoning their assets should encourage them to adopt pollution-prevention measures. At the same time, governments need to acknowledge that they have the responsibility for liability for damage in space and to hold their citizenry accountable through laws.

Do you see any possibility that private actors will solve debris-related issues before governments' intervention?

If we think about the bottom-up approach as companies self-imposing sustainable measures like the extended producer responsibility, namely extending their responsibility on their products to the post-consumer phase to ensure their sustainable disposal, that is certainly something for which I advocate. People should make their satellites reusable and recyclable, as this might show to governments that doing something is possible.

Some insurance companies are looking at debris as a potential market. Do you envision any role for them?

Yeah, of course, I think the role of insurance is critical for space sustainability. Nonetheless, they still lack a robust rationale that might justify their services and products. There is an absence of sustainability metrics and proof of damage caused by debris. If somebody has an anomaly on orbit, how do we know who did it? Was it caused by human made debris? Was it caused by Mother Nature? Currently, there is not enough continuing supervision. Where is the independent evidence that is monitoring who is doing what to whom?

You mentioned the concept of circularity. There are some studies about reusing space debris. Do you see any credible business model for this?

No, there is no marketplace for it because governments do not support it as much as they talk about it. First, using tax credits, governments should encourage satellite operators to recycle satellites (starting from the adoption of circular design techniques) or to delegate third-party organizations to properly dispose of their space assets. This means that if a company (A) does not want to do controlled re-entries and to sign up to an extended producer responsibility as part of the licensing and authorization process in the United States, a servicing company (B) could become a producer-responsibility organization and take care of A's debris. A could delegate that responsibility to B, and this would formally allow company B to rendezvous, remove and deorbit the space debris. This could be a very lucrative and awesome business. Secondly, governments should take the responsibility of developing sustainability metrics, as these might quantify the benefit or detriment of debris to the environment and to businesses, incentivizing monetization. In my opinion, governments have been lethargic and apathetic in moving towards this.

Do you think that some technologies are more needed than others to prevent space objects' collisions?

In space there are all sorts of objects, and one technology cannot solve every issue. An ensemble of technologies is needed, instead.

What about space objects that cannot be recognized as property of any specific satellite operator?

This is where I would try to motivate countries to pass laws. Space laws could be developed along the lines of maritime salvage, where pure salvage is the ability for anyone to clean the ocean and get compensated for it even if it was not performed under a contract. Equivalent laws for space might be very beneficial.

This sounds like something that requires international agreements. Do you think that the states are ready?

This is the objective, putting everything to happen at the United Nations' level. Unfortunately, it is just not going to happen. Since 2019, more than ninety countries have signed for adopting the 21 *Guidelines for the Long-term Sustainability of Outer Space Activities* issued by the United Nations Committee on the Peaceful Uses of Outer Space (UN COPUOS). Have you seen anything that shows how those countries that signed by consensus have done something tangible? Do you see any evidence of how they are implementing these things? In my opinion, the United Nations has difficulties in monitoring this. The *Convention on Registration of Objects Launched into Outer Space* states that countries should register their objects to the United Nations' Register as soon as it is practicable. However, looking at all the registrations since 1984, I found that there are many countries that register their space objects up to five years after their launch. I am pretty sure that "as soon as it is practicable" does not extend to five years. Furthermore, there are companies that never register their space objects. It is surely difficult to manage what you do not know and do not measure. So, it all boils down to gathering evidence and making it public, transparent without being accusatory.

Some think that mandatory regulations are not needed. Do you agree with this?

We have laws about how to behave on the roads, how to behave on the ocean, how to behave in the air. We do not let people decide whatever they want to do. You need to follow these laws and if you do not, there is going to be a consequence on your behavior. That is how we do everything. I do not understand why space should be treated differently. The whole point of UN COPUOS's *Guidelines* was to provide countries with principles to be enforced as national laws.

Is there the possibility that hard law would prevent the space market from growing? What is the right balance between regulating and fostering commercial activities?

For me, laws are not preventing people from making money. Rather, I would say the opposite, laws help people make money. I just think that this fear of laws hindering business traces back to the rhetoric from people that are thriving in a landscape of ambiguity and confusion. Probably, they believe that it is in their best interest to keep things as nebulous as possible, but we are seeing evidence that this situation in the long-term will be unsustainable, and they will be negatively affected as well. If everybody continues their activities without following good conduct, the risk is to lose the space as a resource for humanity.

How long might it take to make people aware and empathetic towards space environment preservation?

I am going to stand up for this every day. I am focused on connecting space with land, air and ocean so that when people talk about environmentalism it flows from ground to space, all as parts of the same whole, Gaia. Earth as a system of systems. That is what I am trying to do, recruiting people and asking them to use the same words of environmentalism of land, air, and ocean. It is not needed to invent new terms. This might even create more confusion. That is why I use the term circular economy. There are not space terms, but waste management terms.

How would you respond to phrases like, "Who cares?", "Why does it matter?", "There are no living beings in space"?

I think a moral and philosophical response would not resonate with most people asking me "how does it affect my life and what is the economic impact?". My goal is to present concrete evidence. I want to show to humanity the interconnectedness of space with all other things and how space influences everyday life.

Norma
Fr

tive

amework

If this is, indeed, a Golden Age of space exploration, it remains to be seen whether we are in the dawn of that age, its *dénouement*, or somewhere in the middle.

If the revolution in commercial space exploration comes crashing down, governments and industry would do well to reflect on what went right, what went wrong, and how might the space programs of the world better organize themselves next time—for there will be a "next time". If the age is somewhere in the middle, with the historical financial and physical barrier of launch at last overcome (or at least, heavily mitigated) by way of reusable rocketry, as humankind makes real efforts at establishing a permanent multiplanetary presence, this is the time to make some hard choices about what we want that presence to look like—without derailing the entire enterprise.

If, however, this is only the dawn of a new age of exploration, this is perhaps the most critical, but also the most delicate, time to figure out what the norms of this new space age should be, in political, economic, scientific, and cultural terms. And indeed those terms are all relevant to space exploration because the "value" of space crosses all areas of life on Earth, including our way of communicating, our economies, the weather we encounter and the climates we endure. Even our philosophies and spiritual outlook are left changed in meaningful ways. Establishing a framework to protect and nurture the growth of space-based infrastructure means that "easy" access to space—still challenging, but without the added challenges of catastrophic space debris or a poisoned planet Earth left behind—will remain available across generational lines. As such, it will mean that who we are as a species is always up for renegotiation. Our incursions thus far into the stars have been an uneasy negotiation of our best selves with our worse impulses.

NASA and Roscosmos exist today because two superpowers wished to demonstrate their abilities to annihilate each other by way of total thermonuclear warfare: a one-button solution to the issue of human existence. And yet both agencies have managed to inspire not because their rockets were the obvious best solutions to lobbing warheads into major metropolitan areas, but because brave men and women strapped themselves onto those missiles to prove that humankind need not be tethered to our planet of origin.

By way of the Apollo program, some of the greatest engineering minds to ever live converged on a cause that served "to organize and measure the best of our energies and skills", as President John F. Kennedy said.[62] And men of uncommon courage volunteered—indeed, lobbied hard—to lead the most objectively terrifying task of exploration ever undertaken by the human race.

No one's hands are clean in space exploration—both agencies participate in space-based military and spy gathering activities—but by and large, both have endeavored institutionally to transcend all that, to treat them not as *raisons d'*être, but rather, the ugly price of doing business, and the public recognizes that, and the public loves them for it. Those agencies, and ESA, the Japanese space agency, the Indian Space Research Organisation, and others, put humanity's best foot forward each time they launch someone into space. It could have all gone so wrong, and yet somehow, despite the efforts of nefarious powers in governments the world over, they project the best of us, and demonstrate a future that might yet be.

But our increasing celestial footprint brings with it new, never before faced challenges, to include the crowding of orbits, increased debris, and potential kinetic conflict in an metaphorical ocean that heretofore has only known peace.

Spacecraft orbiting the Earth between 700 and 900 kilometers in altitude are especially at risk. This is the zone used primarily by Earth observation satellites that collect valuable information about the health of the planet on which we now live. Crowding, collision, and debris issues threaten the safety of the operational environment. Simultaneously, the militarization of space means satellites might be considered legitimate targets during wartime, exacerbating the debris issue, perhaps irrevocably.

Ultimately, governments must work proactively together, and with industry, to develop norms and ideals that govern our relationships *in* space and relationship *with* space, in times of war and peace. To preserve the potentially fragile progress now being made in the commercial sector, all stakeholders, including industry, developing countries, militaries, civil society, and scientists, should be involved in a sustainability framework. And ultimately, every nation is indeed a stakeholder, regardless of whether they have a space program or launch capability. The heavens belong to every creature of the Earth who has ever lived or will ever live, and part of building a true normative framework for space sustainability means making sure everyone involved, directly or indirectly, are cognizant of the role of space in their development and security, and accordingly, why sustainability must be striven for. Collaborative, equitable regulation, if they strike a careful balance, can maintain stability and predictability in the sector. If they strike a careful balance, they can further the incursion of our better angels to the stars above.

If, however, regulations tip too far in the direction of being burdensome, they might be exclusionary to would-be corporate and national participants.

An open, stable space sector, regulated lightly but meaningfully, is attainable. With it will come new paradigms in human interaction. With those interactions, new knowledge, insights, and perspectives will yield new methods of solving problems—not only in space, but down here on Earth as well.

The signatories of the Artemis Accords are an impressive roster of nations sometimes antithetical to one another, but somehow joined on this impossible endeavor: a peaceful human presence on the lunar surface, and then: Mars. The will is there to do the impossible, together. Now is the time to figure out what our "best foot forward" should look like.

A Conversation with

Cynda Collins Arsenault

—

Victoria Samson

CYNDA COLLINS ARSENAULT is co-founder and President of Secure World Foundation (SWF), which works with governments, industry, international organizations, and civil society to develop and promote ideas and actions to achieve the secure, sustainable, and peaceful uses of outer space.

VICTORIA SAMSON is the Washington Office Director for SWF. She works on space security and stability issues, and strives to incorporate international perspectives with the Washington policy debate, and clarify US government and industry strategy for stakeholders outside of the US.

Could you share with us when and how your interest in space sustainability was born?

CCA: I came to the world of space sustainability in an indirect manner. In high school I lost friends to the Viet Nam War. In college I joined the anti-war movement as I saw the damage and post-traumatic stress disorder of returning veterans and the increasing death toll both during the war and after the war by suicide and drug use. As my husband became successful in business, we set up a foundation with the long-term vision of a world beyond war. Space came to my attention as a potential trigger for conflict on Earth or that conflict on Earth could escalate to conflict in space.

VS: I arrived at space sustainability after having worked on military space issues. My first job out of grad school (where I studied international relations) was working for a defense contractor on war-gaming scenarios for what is now the US Missile Defense Agency. After a few years of that, I switched teams—went into the light, as I like to describe it—and started working on arms control issues. At that point, the United States was very casually looking into research on space-based interceptors as a possibility for its missile defense system, so I started focusing on space as an off-shoot of my work monitoring missile defense developments.

What are outer space use's consequences for space sustainability? What are the challenges?

CCA: Our current uses and benefits of space are innumerable—communications, finance, weather, environmental knowledge, tele-health, tele-education, economics, etc. Putting systems into place that can help ensure the sustainable use of space will preserve those benefits for our children and our children's children. New benefits arrive every day—the overview effect, commercial uses, increased knowledge of our planet and the universe.

VS: If we cannot figure out how to promote the long-term sustainable use of space, we risk several things. Some orbits may become too cluttered with space debris or even active satellites to be able to operate in or through safely, or those orbits may become too costly for most to operate in. If counterspace capabilities continue to proliferate and become normalized for usage, it is possible that conflict on Earth could extend to space (or vice versa).

Which regulatory developments are crucial to ensure space sustainability?

CCA: We need to have a general understanding and guidance on space traffic management, general rules of the road, debris mitigation guidelines, on-orbit servicing procedures and policies, shared Space Situational Awareness (SSA) data, the understanding and acceptance of guidelines for best practices, methods of monitoring and enforcement, transparency and communications systems to prevent misunderstandings, and structures for resolving disputes and conflict.

VS: In addition to what Cynda has mentioned, I would include regulations looking at the specific stresses that very large constellations put on the existing space governance system. How do we ensure that certain orbits are not essentially appropriated by a single actor? How do we ensure that satellite operators can communicate with each other in a manner that allows for decisions to be made in a timely manner? What is (if any) the effect of launching/deorbiting thousands of satellites on the Earth's atmosphere and environment? Are cybersecurity best practices being baked into the satellite constellations from the very beginning?

Should current treaties (such as the *Outer Space Treaty*) be adjusted in front of the sector's rapid evolution and expansion?

VS: The *Outer Space Treaty* (OST) has good bones and has formed an excellent framework to provide guiding principles for how the international community approaches and acts in space to date. The space domain is changing but largely what needs to be done is to see how the OST principles apply to the new activities in space; granted, this will take a lot of discussion in order to ensure that this interpretation is done in an equitable manner.

CCA: A legal system often follows rather than proceeds the development of norms and ideals. There are numerous opportunities to create the foundation for governance mechanisms of space—Group of Government Experts (GGE)'s, Working Groups, Track 1 ½ and 2 dialogues, Committee on the Peaceful Uses of Outer Space (UN COPUOS), conferences, and trade meetings. Whatever means, it is important that the work is inclusive of all stakeholders—commercial, developing countries, militaries, civil society, scientists, etc.

Do you think that any mandatory regulations should be enforced to strengthen the currently available voluntary guidelines framework?

VS: When talking about international relations and governance, it is more important to determine if the governance is appropriate to the problem it is trying to solve and if it is to countries' benefit to follow that governance. If it is not to their benefit, the governance is not helpful because countries will not follow it, no matter its format. So, the question becomes, "what is actually needed in terms of the guidelines?".

For example, the "25 years rule" aims to direct satellite operators to reduce the length of time their spacecraft spend in the protected region after the end of their mission, with the guidance being that twenty-five years after the satellites have reached their end of lives, they will either be deorbited into the Earth's atmosphere so they burn up or put in a graveyard orbit out of the way of other satellites. So, the question is, "is that still useful?". In September 2022, the United States Federal Communications Commission (FCC) adopted new rules requiring satellite operators in low-Earth orbit to dispose of their satellites within five years of completing their missions. Is that a useful number?

I think the environment is changing in terms of how countries prioritize this and how they look at what is important themselves. It is becoming more and more relevant for an increasing number of actors. So, I do have hope that we can come to some sort of solution and have ideas about what is going to be useful for these guidelines and principles without necessarily going into opening up the *Outer Space Treaty*.

In the context of the current democratization of the space sector, can regulation and competition go together?

CCA: Most of the world has developed a societal construct based on competition. Yet competition is not possible without rules and order. Even small children make up rules for the games they play. Looking at history, we can see that if the rules are not fair, unrest and violence develop. A key to developing effective rules and regulation is through an inclusive process, where those impacted by the rules are a part of creating them. This includes not only current actors but those that will be impacted in the future.

VS: In addition, there is a fine line to be walked between providing enough regulation so that space is a stable, predictable domain but not so much as to unduly squelch competition. This is a hard balance to hit but an important goal to strive for. As well, it is important to ensure that new and emerging space actors do not have too large of a burden placed upon them, so it is important to include all stakeholders when having these discussions in multilateral fora on issues related to space sustainability.

Do you see any geopolitical barriers in order to guarantee space sustainability? Might geopolitical tensions hinder the preservation of space?

CCA: Space is a complicated environment with complicated issues. Resolving these issues will take focus and creative thought. It will be important to include all actors in the conversation—governments, commercial space, scientists, developing countries, civil society, and beyond that, the general population—men and women who all stand to lose if we are not able to guarantee space sustainability. Creating the spaces for those conversations to happen now will help develop the solutions we need. In addition, by bringing together the various voices and creating cooperative activities in space, we may also help the geopolitical tensions here on Earth.

VS: As always, hard to beat Cynda's eloquence on this! I will point out that while geopolitical tensions have the potential to interfere with space sustainability efforts, they are not necessarily destined to. The 21 *Guidelines for the Long-term Sustainability of Outer Space Activities* that were agreed upon by COPUOS in 2019 happened when that organization had 92 member states, including those with disparate outlooks like the United States, Russia, China, and Iran. Despite that, COPUOS was able to unanimously agree to these guidelines. If space sustainability is made a priority by the global community, progress can occur.

Do you think that war between Russia and Ukraine can change the cooperation scenario?

VS: The actual Space Station is an interesting case. We have put humans in orbit continuously since 2000 and it has been done even through a lot of ebbs and flows at international level. I mean the Space Station is still going, it is still working. It was deliberately built so that both sides—the US side and the Russian side—are needed for the overall spacecraft to function. Now the question is "how much longer will we keep on going?", and that is where the Russians have been making noises about potentially stopping their participation short of what the United States would like. I mean, the US wants to extend the Space Station from the end of 2024, which was the original plan, through 2030. Not all of the other partners have said that they officially support this extension. Russia has said that it will eventually build its own space station but has not said when or even if it will officially stop participating in the International Space Station. I think that a lot can change between now and the end of 2025, which is when this could actually happen. So, I think Russians even recognize that there is still value in the International Space

Station. I think they are trying to walk a fine line between being able to demonstrate their independence and showing their own way without cutting themselves off entirely from the International Space Station because they are getting benefit from that.

CCA: There is always the possibility that conflict on Earth will escalate into conflict in space which is one of the reasons why we founded the Secure World Foundation. The conflict between Russia and Ukraine provides a strong demonstration of the need for rules of the road for activities in space. Just as the use of nuclear weapons is a lose/lose option, so too, is the destruction of satellites in space. Developing a cooperative approach to preserving space for peaceful purposes will serve to benefit all countries.

Concl

usions

Relatively speaking, life appeared on Earth almost instantaneously after its formation, the circumstances absurdly unlikely: a tiny molten ball circling a youthful star, a chance cataclysmic collision that might indeed have destroyed the nascent world, and a consequent atmosphere of metallic soup and iron rain.[63] It took another three billion years, more or less, for multicellular life to appear—likely as a survival mechanism, so great were the single-celled predators, so common were the single-celled prey.[64] So intense was the predation that eventually creatures crawled from the sea—anything was better than perilous life in the deep—and found a way to adapt to the surface and atmosphere. It took over four billion years from the dawn of the Earth for dinosaurs to appear, and for almost two hundred million years, they lorded over the planet. Then, sixty-six million years ago, an asteroid a few miles across slammed into the Earth, and the era of the dinosaurs came to a close. Homo sapiens have only been around for three hundred thousand years, and for two-thirds of that time, it lacked the brain power to do the things we do today.

It is hard to describe the human adventure thus far without using the word "miraculous". None of this should exist: Human civilization, space exploration, this book. None of it.

And yet it does, and from the perspective of time and history, the problems we now face are grand and glorious—not problems at all, but *opportunities*. Against all odds, here we are, and if the story of life offers any moral or solace, it is that the creatures of the Earth are nothing if not adaptable. We rise to the occasion. And we endure.

The space sustainability issues explored in this book are but a sampling of those that will drive the next century of exploration, and indeed, touch the face of the human experience going forward. Each in its own way gnaws at certain fundamental questions that must be answered. To whom does space belong? Presently, the answer as a practical matter is governments, with a surge of activity from private industry—sometimes called "New Space"—now pushing and sometimes outpacing ossified state structures. The simple fact that these issues are not only worthy of discussion, but essential to explore, is a triumph: declaration after declaration of how far we have come.

In considering the issues, care must be taken to sustain the future of space exploration without halting its present. Sweeping declarations and iron-fisted desires for immediate results are certain to be wrong. Carved outside the Temple of Apollo at Delphi is the statement translated as "Certainty Brings Insanity". Few better words apply here, and true space sustainability will require careful negotiation among peoples and nations of the world. The United Nations Sustainable Development Goals are thus ideally suited to bring about desirable outcomes in space issues, as they offer a direction and rationale without draconian imposition. Industry should not be hindered, but rather, should be incentivized to respond to clear and shared rules.

Terrestrially speaking, we already know that such rules, agreed upon by the world's governments, with standardized technologies and procedures, are possible, and indeed foster innovation while keeping humans and hardware safe. The International Civil Aviation Organization (ICAO) of the United Nations coordinates the protocols, infrastructure requirements, and recommended practices for global aviation. A plane can take off safely from the United States and land smoothly in China, which, given the complications of international relations, let alone the complexity of an average airliner, is an achievement all the more impressive. That the first iteration of the ICAO stood up in 1919—a mere sixteen years after the Wright Flyer, and eight years before Charles Lindbergh flew across the Atlantic—demonstrates how rapidly governments can come to consensus on risky and highly technical endeavors.[65] This offers one example of a United Nations-driven regulatory environment that effectively threads the same needle with which space travel is now faced.

In practice, the International Space Station is the best case thus far undertaken for how international cooperation even in uncertain times can facilitate the building of something of the grandeur of the Great Pyramids. It is certainly the greatest engineering achievement of the last fifty years.

The whole history of the station is fraught with political tensions that have at turns threatened to leave the orbital laboratory uncrewed (or at least, crewed only by Russians), defunded, and even disassembled in orbit during international tiffs. And yet it has endured. The station is by and large funded by the United States (indeed, in a first, the United States paid the Russians to build their modules in the 1990s), and once NASA moves on to the next thing—presently construction of an orbital outpost around the Moon—it remains to be seen whether the International Space Station will survive or be dropped into the ocean.[66]

There seems to be mixed interest in a commercial assumption of International Space Station activities beyond tourism. Space manufacturing is highly specific work, and the prospect of smaller, tailor-made space stations obviates the renovation of the station for private work. It might be the case that it is simply cheaper and easier to start afresh than renovate, however good the bones of the station might be. Even if the space station does not survive the Gateway station in cis-lunar space, or a base on the Moon itself, the lessons learned simply in launching its various modules, assembling them with international crews, and keeping them aloft for decades, will endure and offer direct guidance in space sustainability—itself an international endeavor by necessity. That the station is operated by five space programs collaborating in wartime and peace, and crewed by more than 250 astronauts, taikonauts, and cosmonauts from twenty different countries, certainly drives home the point that international comity in space is possible.[67] In short, regulation is not inherently destructive, and bringing the world together for common goals in space is more common than some might realize.

The Artemis Accords, first signed in 2020, with other countries such as Nigeria and Rwanda joining as recently as December 2022, introduce sustainability into a major international endeavor in space.[68] Presently there are dozens of signatories to the agreement, which builds on the United Nations' *Outer Space Treaty* of 1967. Specifically, among other things, the Artemis Accords commit signatories "to plan for the mitigation of orbital debris, including the safe, timely, and efficient passivation and disposal of spacecraft at the end of their missions, when appropriate, as part of their mission planning process".[69]

Moreover the accords affirm the *Outer Space Treaty*'s mandate that all objects launched into space must be registered, and sets into place an initial framework for international Artemis collaborations to assign registrars. This is significant because signatories are bound to the agreement; any nation that ignores said sustainability guidelines risks expulsion from the program. It is the first solution to the space debris problem that punishes misbehavior.

Regulation is balanced with enticement. The Artemis Accords carve a place for the extraction by individual signatories of resources on celestial objects. The accords require signatories to "affirm that the extraction of space resources does not inherently constitute national appropriation under Article II of the *Outer Space Treaty*, and that contracts and other legal instruments relating to space resources should be consistent with that treaty".

Article II states: "Outer space, including the Moon and other celestial bodies, is not subject to national appropriation by claim of sovereignty, by means of use or occupation, or by any other means".[70] Previously, there was some question as to whether elements harvested from the Moon and elsewhere constitute appropriation.

There is another United Nations document that addresses the subject: The *UN Agreement Governing the Activities of States on the Moon and Other Celestial Bodies*. It states: "Neither the surface nor the subsurface of the Moon, nor any part thereof or natural resources in place, shall become property of any State, international intergovernmental or non-governmental organization, national organization or non-governmental entity or of any natural person".[71] No countries capable of launching humans into space have ratified the latter agreement.

The political momentum would seem to suggest that space mining is an inevitability as an inducement for independent commercial investment unsubsidized by nation-states. (Commercial interest is necessary because nation-states have demonstrated neither the will nor ability to fund major, long-term industrial lunar or Martian activities on timelines sufficient for meaningful progress.) By giving nations this "out" from any ambiguity in the *Outer Space Treaty*, it rewards sustainability with the prospect of mining resources on the Moon. In addition, the Artemis Accords represent an example of the international community establishing clear rules. In doing so, they enable New Space to advance into a new and exciting sector.

NASA's Artemis program matters with respect to sustainability and global space cooperation because it has proven to be quite successful thus far in attracting international investment at various levels.[72] In other words, not every nation needs to pony up billions for a lunar habitat or extraordinary new rocket. The model allows smaller countries, or those with smaller space programs, to push the Artemis program forward with contributions akin to Canada's "Canadarm" robotic payload arm on the International Space Station—a relatively small but important part of the station with great utility to astronauts on board. Such contributions to the program are avenues through which participating nations might get their own citizens on the lunar surface—precisely the sort of inventive, nimble, essential enticements that win proponents around the world. The catch is that participation in the Artemis program requires nations to adopt the Artemis Accords. This places the United States and NASA squarely in control of the most likely major lunar development program.

Although treaties represent an important regulatory framework, voluntary mitigation strategies have a prominent role as well in the issue of space sustainability. In 2007, the United Nations first endorsed the *Space Debris Mitigation Guidelines* of the United Nations Committee on the Peaceful Uses of Outer Space (UN COPUOS). In 2019, the United Nations published the *Guidelines for the Long-term Sustainability of Outer Space Activities* of the Committee on the Peaceful Uses of Outer Space. These are "living documents", revised as situations warrant. According to the United Nations, such documents are premised "on the understanding that outer space should remain an operationally stable and safe environment that is maintained for peaceful purposes and open for exploration, use and international cooperation by current and future generations, in the interest of all countries, irrespective of their degree of economic or scientific development, without discrimination of any kind and with due regard for the principle of equity".[73]

State space agencies have crucial roles to play in space sustainability, infrastructure development, and the protection of planet Earth. The celestial equivalent of cleaning out old landfills is the removal of space debris from low-Earth orbit. It remains not critical to ongoing operations. As stated previously, the International Space Station must now routinely adjust altitudes to avoid debris collision, a problem made immeasurably worse by the Russian missile strike on KOSMOS 1408.[74]

Russia is not the only country to destroy a satellite in space—the United States, India, and China have, as well—but that particular anti-satellite demonstration created an especially hazardous debris cloud of 1,500 pieces in a region of space where humans work.[75] From a space sustainability standpoint, this is analogous to an intentional Chernobyl reactor core meltdown, with global implications.

In the months immediately following the Russian antisatellite demonstration, the SpaceX Starlink constellation was expected to perform as many as 80 maneuvers per day.[76] The debris field has spread since the demolition, and will continue to do so. Worse yet, though the KOSMOS 1408 debris field can be avoided, and given space "clean up" technologies, might one day be mitigated, future such "demonstrations", or even direct actions, might render even clean-up—already an unproven ability—more difficult.

Satellite constellations themselves are potential sources of debris. Starlink's collision avoidance systems require each Starlink satellite to actually work on launch, which is not always the case. Starlink satellites have seen an overall failure rate of approximately 4.3%, by one estimate, with some launch batches worse than others.[77]

Though each Starlink node is designed to deorbit and disintegrate with zero debris, the possibility exists for extant space debris to impact "dead" nodes before their orbits decay. The worst case scenario of this is the potential for an uncontrollable cascade effect, in which satellites destroy other satellites with inadequate time to avoid spreading debris.

Governments are aware of the predicament. In 2022, the United States Space Force initiated a program called Orbital Prime to kickstart industry development, with awards up to 1.5 million dollars.[78] Japan-based Astroscale launched in 2021 a project called End-of-Life Services by Astroscale-demonstration, or ELSA-d: the first on-orbit debris cleanup demonstrator spacecraft. Last year, OneWeb, a London-based private satellite Internet access provider, offered proof of a potential market for such cleanup services, beyond state funding, when it signed a 3.5 million dollars deal with Astroscale to help deorbit satellites.[79] The United Kingdom Space Agency and European Space Agency contributed €14.8 million to push the technology even further, with a multiple-satellite de-orbiter spacecraft targeted for 2024.

All this is right and just. At risk of gross oversimplification, governments the world over have long funded clean-up efforts both through national environmental programs as well as on a municipal level—garbage trucks haul off refuse every single day. If humankind is indeed destined to live among the stars, our behaviors and activities will follow us. Perhaps the best case scenario for humankind is that going to space as a sanitation worker is a job one does—unpleasant, perhaps, and certainly long and difficult work, but good and dignified work all the same. The greatest success of the space programs of the world would be for jobs in space to be the ones people turn down because of their tedium. And as has been demonstrated for centuries, where there is money to be made, industry will follow.

In the shortest of this list of medium term challenges, the threat of space debris seems the most likely to be addressed.[80] In other words, space debris is an area where there is universal consensus that a solution is needed, and money being spent on solving it. Low-Earth orbit is valuable real estate for a burgeoning economic sector.

In addition to failed nodes causing a destructive cascade effect, satellite constellations also pose space sustainability issues in the area of planetary defense. Specifically, astronomers fear that once broadband constellations achieve critical mass (in the tens of thousands of individual satellites), they will disrupt ground-based wide-field observation telescopes, which are critical for detecting "potentially hazardous objects" such as incoming asteroids and comets that might strike Earth. This problem can be mitigated by space-based detection solutions, but whether such spacecraft as NASA's Near Earth Object Surveyor, to be built by Jet Propulsion Laboratory, launch before such mega-constellations as Starlink or Amazon's Project Kuiper are fully deployed, remains unclear.

But it provides an interesting test case. The nations most helped by Internet-providing satellite mega-constellations are amongst the poorest in the world. In Africa, for example, fifteen nations have an Internet access rate of less than one-quarter of their respective populations. Three nations have rates lower than ten percent.[81] In the developed world, residents of poorer rural areas likewise have little access to the Internet.[82]

Meanwhile, astronomers have begun waging a public war against satellite mega-constellations, arguing that satellite nodes interfere with observation campaigns.[83] Their "save the night sky" initiative has brought a thorny philosophical problem into play. In the interest of space sustainability, should the poorest people in the world, lacking connectivity to the world's digital societies and knowledge base, simply "take one for the team" so that astronomers might get better measurements of the universe nine billion years ago? What is the value of knowledge for knowledge's sake, versus extant information to bring parity to peoples already victim to gross injustice, and with few other demonstrable ways to connect to the wider world? If there is a way to have it both ways, it likely resides in state resources to fund research that might somehow mitigate the telescope-obscuring effects of satellite mega-constellations, while also fostering constellation development to bring online the final millions in the farthest reaches of the Earth. It is certainly a space sustainability conversation with potential to raise greater awareness of these issues among the public at large.

If there is anything else to be said about such constellations, it is that space weather—that is, the caprices of the Sun's corona—poses a hazard to them, and to low Earth orbit. Once mega-constellations reach critical mass, a space weather event could possibly trigger collisions, which, given the close proximity nodes have to one another, could trigger yet more collisions, until, just as might be the case of an anti-satellite missile, low-Earth orbit is devastated with debris. The Space Weather Prediction Center in Boulder, Colorado, and NASA, have monitoring stations and spacecraft to act as warning buoys before a coronal mass ejection might slam into Earth, but these spacecraft are aging. A responsible posture for space sustainability would involve protecting space-based assets in Earth's orbit and its Lagrange points by launching increasingly capable space weather satellites.

The scale of activity in space is nearly as hard for the human mind to grasp as the scale of space itself. From helicopters on Mars to autonomous rovers on the Moon to nuclear reactors in orbit, scientific exploration has never seen greater success, nor more lavish funding. Given that space sustainability is about not only addressing problems now manifesting, but anticipating problems to come, human culture must also be considered as something worthy of preserving.

Consider the likely near-term future for humanity. If indeed the Artemis program is a success, or the Mars ambitions of the private sector, astronauts will be very tempted to revisit the greatest achievements of engineering and human bravery. Who would not want to walk where Neil Armstrong made "one small step for a man, one giant leap for mankind"? But because of the nature of an airless body, the footprints of the Apollo astronauts—barring an unexpected and unfortunate meteor impact event—will remain unchanged for millions of years. Part of sustaining space is sustaining vital cultural heritage sites on other celestial bodies. Space archaeology and anthropology must therefore be part of any conversation on sustainability. This might take the form of a United Nations Educational, Scientific and Cultural Organization (UNESCO)-type organization, or even that organization itself, to preserve and protect humanity's early triumphs off Earth from space tourism, or space-based commercial enterprise. Yet again, the establishment of clear rules by the international community that all must obey would promote sustainability and give a level playing field to governments and private industry.

While it seems unlikely that a company would run a bulldozer across Tranquility Bay, world heritage sites are destroyed all the time through war and carelessness. Never again, for example, will anyone cast their gaze on the Buddhas of Bamiyan, once the tallest statues of Buddha in the world.

It is a near-certainty that humanity will not leave our warlike nature behind when at last we live on other worlds. Obliterating the site of Apollo 15 might be a nice way for one country at war with the United States to in some way retaliate on a visceral level. Such actions are considered a war crime on Earth, and in general (though there are terrible exceptions), warring nations do a passable job of avoiding such actions. This is as good a time as any to establish a framework preventing any such action against historical places and equipment (Mars rovers, for example) elsewhere in the cosmos. If we have reached a point where we can destroy them, it stands to reason that we will have reached a point where we can build museums around them. Space exploration is the most deeply human endeavor ever undertaken. The achievements that got us to the stars in the first place must take priority over our cultivation of the cosmos. Mars is a big planet, and our Moon is a large one indeed. There is plenty of room without needing to disrupt what came long before.

Another unintuitive part of the space sustainability conversation, and grossly underappreciated, is the need to communicate the benefits of space and space programs to the people who oftentimes pay for it, and always use it. Every year, NASA releases a book titled "Spinoff" which explains all the technologies developed for space exploration, whether human or robotic, and the ways those technologies have trickled down, through the agency's Technology Transfer program, into the everyday lives of men and women. Pressure-based low-tire sensors in automobiles were born of the space shuttle, for instance.[84] The full-body plastic sheets runners drape around themselves after a marathon are called, informally, "space blankets", and derive from radiation barriers developed for use on spacecraft.[85] Fogless goggles worn by snow-skiers were born from the need by NASA to keep spacecraft windows clear at launch.[86]

Broadly speaking, the government, industry, academia, and the public, knowingly or not, converge on the same point: space is a public good. No one ever navigated with a GPS and then cursed when it worked. (They have, however, certainly been upset when it did not.) But people, overall, do not follow the doings of NASA, ESA, JAXA, or any other space agency. People are often oblivious to things that seem obvious to those who follow the space program and all its doings—and that is OK.

But the small, integrated improvements in day-to-day lives are important to highlight, and more so is the urgency for space programs, public and private, to remind the masses that the very infrastructure of day-to-day life is enabled by—or dependent upon—continued spending on space-related endeavors, ever-improving infrastructure in low-Earth orbit and beyond, and sustainability to protect all that has been achieved and will yet be achieved.

The danger of space sustainability as a notion is the difficulty of maintaining a careful balance between what must be done and what must not. The peril of living through the Golden Age of any category is the false feeling that it will axiomatically last forever; that it is impervious to destruction; and that it is easily steered.

The way we now explore space is less like a rocket soaring upward, veering this way or that by engines on gimbals, than it is a loaded freight train racing down tracks still being built. The best regulators can hope to do is stay ahead of the railroad engine, shifting the directions of rails a little at a time, this way or that, keeping the whole thing stable and moving forward. Any steep movement in one direction or the other could be catastrophic—and to be sure, catastrophe is inherent in the space enterprise.

The first time astronauts are lost on a private rocket, private station, or private colony, the industry will reel and government overseers will pounce. A full assessment of liability for such an event remains untested in court. If for no other reason, the true durability of presently-robust space exploration enterprises is utterly unknown.

A total embrace of the space sustainability topics outlined in this book will not likely come of their own accord. Perhaps the most effective way of unifying efforts in space sustainability and its goals is, on the outset, by way of an addition to the United Nations Sustainable Development Goals. Presently, the seventeen goals address all areas of the human experience, from starvation to safeguarding cultural heritages. Each of the seventeen goals are interlinked, mimicking the holistic human experience.[87] (Breathing clean air, after all, is a poor consolation when you are starving to death.) Space sustainability is an obvious eighteenth goal because it is both its own domain with its own set of challenges, while also being intimately connected to terrestrial challenges.

Earth-monitoring satellites foster a greener planet by providing government and industry with nearly real-time data on everything from coastlines to farmland, and directly support goals two and three of the sustainability list, namely: ending hunger and promoting wellbeing, as well as thirteen, fourteen, and fifteen, each of which is concerned with the environment and climate change. Broadband Internet mega-constellations likewise bolster goals four (ensuring equitable access to education), eight (promoting sustained economic growth), and sixteen (fostering peaceful and inclusive societies) by way of expanding the access to the world's collective knowledge, and through unprecedented communications across borders and cultures. Space-based power generation can revolutionize goal nine, which requires building resilient infrastructure, including electricity, and goal eleven, which calls for resilient and sustainable cities and communities. Collectively, the areas of space sustainability discussed in this book directly support goal seventeen, which calls for strengthening the means of implementation for the sustainability guidelines.

This is all consistent with the role space plays in our lives. Not as a place beyond Earth's surface, but rather, the place where Earth and her inhabitants are at this very moment— immersed and alive. Space sustainability is, at its core, *sustainability itself*, and as essential to the prosperity and survival of our species as clean water and air. But it will require humans, ever fortunate on the cosmic timeline, beneficiaries to an Earth that nearly wasn't.

We have, as a species, made good choices and bad, and they have taken us far. How much further and farther we go is entirely on us. The path to a cleaner Earth and a secure future will require fortifying the nest, and also leaving it, for the alabaster and apricot horizons of the Moon and Mars. It can be done, and responsibly, but it requires a willingness to prioritize tomorrow over today, and what might be, for what presently is. We are capable of making the right choices. We are capable of sustaining ourselves and our world. And one day, our *many* worlds.

Notes

A NEW PHILOSOPHY FOR SPACE AND SUSTAINABILITY

[1] Longrich, N.R., *When did we become fully human? What fossils and DNA tell us about the evolution of modern intelligence*, "The Conversation", 2020; http://theconversation.com/when-did-we-become-fully-human-what-fossils-and-dna-tell-us-about-the-evolution-of-modern-intelligence-143717

[2] May, A., *The Space Business from Hotels in Orbit to Mining the Moon. How Private Enterprise is Transforming Space*, London: Icon Books Ltd, 2022.

[3] Wellernstein, A., *A bomb without Einstein?*, "Restricted Data. The Nuclear Secrecy Blog", 2014, June 27; https://blog.nuclearsecrecy.com/2014/06/27/bomb-without-einstein/

[4] Brundtland, G., *Report of the World Commission on Environment and Development: Our Common Future*, United Nations General Assembly, document A/42/427, 1987.

[5] Baalke, R., *Mars Meteorite Home Page*; https://www2.jpl.nasa.gov/snc/

[6] Limos, M.A., *There Could Be Dinosaur Bones on the Moon, Says Scientist*, "Esquire", 2021, January 20; https://www.esquiremag.ph/long-reads/features/dinosaur-bones-on-moon-a00293-20210120

[7] Martin, L. and Wilson, N. (Eds.), *The Palgrave Handbook of Creativity at Work*, London: Palgrave Macmillan, 2018; https://doi.org/10.1007/978-3-319-77350-6

[8] The Sustainable Development Goals, issued in 2015 by the United Nations as core element of the 2030 Agenda for Sustainable Development, are a set of 17 objectives aimed at providing the global community with a shared agenda to pursue a future of peace and prosperity for the planet and for humankind. More specifically, they call private and public players to partner to end poverty, improve health and education, minimize inequalities, foster economic development, and preserve the environment. Source: https://sdgs.un.org/goals

[9] Yellowstone National Park, *Frequently Asked Questions: "Bison"*, 2021; https://www.nps.gov/yell/learn/nature/bisonfaq.htm

[10] Camus, A., *The Myth of Sisyphus* (Second Vintage international edition). Vintage International: Vintage Books, a division of Penguin Random House LLC, 2018 (1942).

[11] Nash, E.R., *NASA Ozone Watch: Ozone hole history facts*, NASA, 2018; https://ozonewatch.gsfc.nasa.gov/facts/history_SH.html

[12] *Handbook for the international treaties for the protection of the ozone layer: the Vienna Convention (1985), the Montreal Protocol (1987)*, Nairobi, Kenya: Ozone Secretariat, United Nations Environment Programme, 2000.

[13] Galileo Galilei, *Discourses and Mathematical Demonstrations Concerning Two New Sciences*, Hassell Street Press, 2021 (1638).

[14] Loyson, P., *Galilean Thermometer Not So Galilean*, "Journal of Chemical Education", 89 (9), 2012, July 27; https://doi.org/10.1021/ed200793g

[15] Blumenthal, I., *The development of the clinical thermometer*, "Proceedings of the Royal College of Physicians of Edinburgh", 28 (1), 1998.

[16] Kellett, J., *The Assessment and Interpretation of Vital Signs*, in M.A. DeVita, K. Hillman, R. Bellomo, M. Odell, D.A. Jones, B.D. Winters, G.K. Lighthall (Eds.), *Textbook of Rapid Response Systems*, New York: Springer International Publishing, 2017; https://doi.org/10.1007/978-3-319-39391-9_8

[17] Churpek, M.M., Adhikari, R., Edelson, D.P., *The value of vital sign trends for detecting clinical deterioration on the wards*, "Resuscitation", 102, 2016; https://doi.org/10.1016/j.resuscitation.2016.02.005

[18] New Mexico Museum of Space History, *Galileo Galilei*, 1991; https://www.nmspacemuseum.org/inductee/galileo-galilei/

[19] *How satellites are used to monitor climate change*, "Carbon Brief", 2016, February 18; https://www.carbonbrief.org/interactive-satellites-used-monitor-climate-change

[20] *Apollo 1*, NASA, 2017, August 7; http://www.nasa.gov/mission_pages/apollo/missions/apollo1.html

[21] In 2017, the United States started the Artemis program, a project intended to return humans to the surface of the Moon, with the aim of taking a step forwards towards the exploration of other celestial bodies (starting from Mars). The Artemis Accords consist in a series of multilateral agreements between the United States and other countries intended to establish the signatories' collaboration in the context of the Artemis program according to principles grounded in the *Outer Space Treaty* (guidelines published in 1967 by the Unit-

ed Nations; they are the main international treaty that governs activities related to space exploration and use). Source: https://www.nasa.gov/specials/artemis-accords/index.html

[22] Orloff, R.W., *Apollo by the Numbers: A Statistical Reference*, National Aeronautics and Space Administration, 2013.

[23] Brown, D., *SpaceX's Launch Control Room: 3 Rocket Missions in 31 Hour*, "The New York Times", 2022, December 29; https://www.nytimes.com/2022/12/29/science/spacex-launch-mission-control.html

[24] Ansar, A., Flyvbjerg, B., *How to Solve Big Problems: Bespoke Versus Platform Strategies*, "Oxford Review of Economic Policy", 610, 2022, 127919.

[25] Crutzen, P., Stoermer, E., *The Anthropocene*, "Global Change Newsletter", 41, 2000.

[26] The Anthropocene Working Group, *The Anthropocene: an update*, "Geo Q. 12", 2014.

[27] Fagan, R., Mordecai, M., *Before Soleimani's death, concerns about Iran had fallen in many countries including the U.S*, Pew Research Center, 2020, January 23.

[28] Wolfe, F., *Russian Direct Ascent ASAT Test Generates More Than 1,500 Pieces of Trackable Debris*, "Defense Daily", 2021, November 15; https://www.defensedaily.com/russian-direct-ascent-asat-test-generates-more-than-1500-pieces-of-trackable-debris/space/

[29] Garcia, A.M., *Space Station Maneuvers to Avoid Orbital Debris*, NASA, 2022, October 24; https://blogs.nasa.gov/spacestation/2022/10/24/space-station-maneuvers-to-avoid-orbital-debris/

SPACE RESEARCH FOR EARTH

[30] Brown, D., *Why Is NASA Neglecting Venus?*, "The Atlantic", 2017, January 19; https://www.theatlantic.com/science/archive/2017/01/venus-lost-generation/513479/

[31] Garthwaite, J., *What other planets can teach us about Earth*, "Stanford News", 2020, March 4; https://news.stanford.edu/2020/03/04/planets-can-teach-us-earth/

SATELLITE SERVICES FOR CLIMATE CHANGE

[32] Gallup, *Environment*, "Gallup", 2022; https://news.gallup.com/poll/1615/Environment.aspx

[33] Holmes, S.A., *Bradley Effect*, "Pollsters Debate", 2008, October 12.

[34] ESA, *What is an Essential Climate Variable?*, ESA Climate Office, 2023; https://climate.esa.int/en/evidence/what-are-ecvs/

SPACE SOLAR POWER

[35] Space Energy Initiative, *Space-Based Energy solutions to address global energy challenges*, 2023; https://spaceenergyinitiative.org.uk/

[36] Xinhua, *China to build space-based solar power station by 2035*, "Xinhuanet", 2019, December 2; http://www.xinhuanet.com/english/2019-12/02/c_138599015.htm

[37] Snowden, S., *Solar Power Experiment Launched by Navy Research Lab On X-37B Space Plane*, "Forbes", 2020, May 27; https://www.forbes.com/sites/scottsnowden/2020/05/27/solar-power-experiment-launches-on-secret-space-plane/

[38] Arizona Power Authority, *History of Hoover Dam*, 2022; https://powerauthority.org/about-us/history-of-hoover

[39] Ramirez, R., *The West's historic drought is threatening hydropower at Hoover Dam*, "CNN", 2022, August 16; https://www.cnn.com/2022/08/16/us/hoover-dam-hydropower-drought-climate/index.html

COMMUNICATING AND MOVING ON THE MOON

[40] Davies, C., *ESA pushes ahead on Starlink-GPS style hybrid network for the Moon*, "SlashGear", 2021, May 20; https://www.slashgear.com/esa-pushes-ahead-on-starlink-gps-style-hybrid-network-for-the-moon-20673779/

[41] *Future development of lunar economy planned by Telespazio and Inmarsat*, "Telespazio", 2022, July 18; https://www.telespazio.com/en/press-release-detail/-/detail/pr-moonlight-inmarsat

[42] The Lunar Gateway is a key component of the Artemis program consisting in a multi-purpose station orbiting the Moon and serving as an outpost both to support human presence on the Moon and for exploration missions directed towards other celestial bodies. Source: https://www.nasa.gov/gateway

[43] Papike, J.J., Simon, S.B., Laul, J.C., *The lunar regolith: Chemistry, mineralogy, and petrology*, "Reviews of Geophysics", 20 (4), 1982; https://doi.org/10.1029/RG020i004p00761

[44] Li, S., Lucey, P.G., Milliken, R.E., Hayne, P.O., Fisher, E., Williams, J.-P., Hurley, D.M., Elphic, R.C., *Direct evidence of surface exposed water ice in the lunar polar regions*, "Proceedings of the National Academy of Sciences", 115 (36), 2018; https://doi.org/10.1073/pnas.1802345115

[45] *Mars & Beyond. The road to make humanity multiplanetary*, "SpaceX"; https://www.spacex.com/human-spaceflight/mars/

SPACE HABITATS CIRCULAR TECHNOLOGIES

[46] Roa, J., *Introduction. Current challenges in space exploration*, in Id., *Regularization in Orbital Mechanics*, Berlin: De Gruyter, 2017; https://doi.org/10.1515/9783110559125-001

[47] Zabel, P., Schubert, D., Bamsey, M., *Eden ISS. Executive Summary 12/2014*, Zenodo, 2014; https://doi.org/10.5281/zenodo.29670

SPACE MINING TECHNOLOGIES

[48] Brown, A., *As Unpredictable as Humans:* I, Robot *by Isaac Asimov*, "Tor.Com", 2022, June 29; https://www.tor.com/2022/06/29/as-unpredictable-as-humans-i-robot-by-isaac-asimov/; Lufkin, B., *Asteroid Mining in Fiction, Past and Present*, "Livescience", 2012, April 24; https://www.livescience.com/19862-asteroid-mining-fiction-present.html

[49] Nozette, S. (Ed.), *Defense Applications of Near-Earth Resources*, La Jolla: California Space Institute, 1983; https://apps.dtic.mil/dtic/tr/fulltext/u2/a340021.pdf

[50] The *Outer Space Treaty* (*Treaty on Principles Governing the Activities of States in the Exploration and Use of Outer Space, including the Moon and*

Other Celestial Bodies) was promoted and opened for signature by the Russian Federation, the United Kingdom and the United States in January 1967 and came into force in the autumn of that year, under the auspices of the United Nations. Today, as then, it provides the basic framework of international space law. As of December 2022, there were 112 parties and 89 signatories. Source: https://treaties.unoda.org/t/outer_space

SPACE SECTOR SUSTAINABILITY

51 Berry, D.S. *et al.*, *Automated Spacecraft Conjunction Assessment at Mars and the Moon - A Five Year Update*, NASA, 2018; https://hdl.handle.net/2014/48222

52 The International Telecommunication Union (ITU) defines HAPS (high-altitude platform stations) as radio stations located on an object at an altitude of 20-50 kilometers and at a specified, nominal, fixed point relative to the Earth, intended to provide broadband connectivity in regions with limited or no terrestrial networks, such as remote areas or in areas struck by natural disasters. Source: https://www.itu.int/en/mediacentre/backgrounders/Pages/High-altitude-platform-systems.aspx

53 In 2019, the UN Committee on the Peaceful Uses of Outer Space published 21 *Guidelines for the Long-term Sustainability of Outer Space Activities*, which were undersigned by 92 member states (including the United States, Russia, China and Iran), with the purpose of preparing a set of measures recognized internationally and aimed at ensuring the long-term sustainability of space activities beyond the Earth's atmosphere and promoting safety in space operations. Source: https://www.unoosa.org/res/oosadoc/data/documents/2018/aac_1052018crp/aac_1052018crp_20_0_html/AC105_2018_CRP20E.pdf

SPACE DEBRIS

54 Liou, J.-C., *Risks from Orbital Debris and Space Situational Awareness*, Washington DC: IAA Conference on Space Situational, 2020; https://ntrs.nasa.gov/api/citations/20200000450/downloads/20200000450.pdf

55 United States Department of State, *The Montreal Protocol on Substances That Deplete the Ozone Layer*, 2023; https://www.state.gov/key-topics-office-of-environmental-quality-and-transboundary-issues/the-montreal-protocol-on-substances-that-deplete-the-ozone-layer/

[56] IEA, *The Covid pandemic has slowed progress towards sustainable energy goals even as renewables continue to gain ground*, 2022, April 1; https://www.iea.org/news/the-covid-pandemic-has-slowed-progress-towards-sustainable-energy-goals-even-as-renewables-continue-to-gain-ground

[57] Jakhu, R.S., Jasani, B., McDowell, J.C, *Critical issues related to registration of space objects and transparency of space activities*, "Acta Astronautica", 143, 2018; https://doi.org/10.1016/j.actaastro.2017.11.042

[58] Roberts, T.G., *Space launch to Low Earth Orbit: how much does it costs?*, "Aerospace Security", 2022; https://aerospace.csis.org/data/space-launch-to-low-earth-orbit-how-much-does-it-cost/

[59] Tomaswick, A., *According to a US Auditor, Each Launch of the Space Launch System Will Cost an "Unsustainable" $4.1 Billion*, "Universe Today", 2022, March 14.

[60] ESA, *OMAR. On orbit manufactured spacecraft*, 2022, June 5; https://nebula.esa.int/content/omar-%E2%80%93-orbit-manufactured-spacecraft

[61] Harbaugh, J., *NASA's Centennial Challenges: 3D-Printed Habitat Challenge*, NASA, 2016, October 5; http://www.nasa.gov/directorates/space-tech/centennial_challenges/3DPHab/about.html

NORMATIVE FRAMEWORK

[62] Kennedy, J.F., *Historic speeches: Rice University, on the Nation's Space Effort*, 1962, September 12; https://www.jfklibrary.org/learn/about-jfk/historic-speeches/address-at-rice-university-on-the-nations-space-effort

CONCLUSIONS

[63] Service, R., *How an ancient cataclysm may have jump-started life on Earth*, "Science", 2019; https://doi.org/10.1126/science.aaw6068

[64] *Ancient origins of multicellular life*, "Nature", 533, 441, 2016; https://doi.org/10.1038/533441b; Herron, M.D., Borin, J.M., Boswell, J.C., Walker, J., Chen, I.-C. K., Knox, C.A., Boyd, M., Rosenzweig, F., Ratcliff, W.C., *De novo origins of multicellularity in response to predation*, "Scientific Reports", 9 (1), 2019; https://doi.org/10.1038/s41598-019-39558-8

[65] *Convention Relating to the Regulation of Aerial Navigation*, signed at Paris, 1919, October 13.

[66] Brown, D.W., *The Mission*, Custom House – Harper Collins, New York 2021.

[67] Howell, E., *International Space Station: everything you need to know*, "Space.com", 2022, August 24; https://www.space.com/16748-international-space-station.html

[68] Warner, C., *NASA Welcomes Nigeria, Rwanda as Newest Artemis Accords Signatories*, NASA, 2022, December 12; http://www.nasa.gov/feature/nasa-welcomes-nigeria-rwanda-as-newest-artemis-accords-signatories

[69] United States Department of State, *Artemis Accords: Principles for Cooperation in the Civil Exploration and Use of The Moon, Mars, Comets, and Asteroids for Peaceful Purposes*, 2020, October 13.

[70] United Nations, *Treaty on Principles Governing the Activities of States in the Exploration and Use of Outer Space, including the Moon and Other Celestial Bodies*, 1967.

[71] United Nations, *UN Agreement Governing the Activities of States on the Moon and Other Celestial Bodies*, 1979.

[72] NASA Office of the Inspector General, *Artemis Status Update*, 2021.

[73] UN Committee on the Peaceful Uses of Outer Space, *Guidelines for the Long-term Sustainability of Outer Space Activities*, 2021; https://spacesustainability.unoosa.org/content/the_guidelines

[74] Brokaw, S., *International Space Station maneuvered around Russian satellite debris*, "UPI", 2022, June 20; https://www.upi.com/Science_News/2022/06/20/international-space-station-maneuvers-around-russian-satelitte-cosmos-1408-debris-nasa/2891655748476/

[75] Corrao, G., *Analysis of the Cosmos 1408 ASAT*, "Linkedin", 2022, March 18; https://www.linkedin.com/pulse/analysis-cosmos-1408-asat-giuseppe-corrao/

[76] Foust, J., *Russian ASAT debris creating "squalls" of close approaches with satellites*, "SpaceNews", 2022, February 18; https://spacenews.com/russian-asat-debris-creating-squalls-of-close-approaches-with-satellites/

[77] Todd, D., *One Starlink launch batch does not look very healthy: one in five satellites appear to have failed*, "Seradata", 2022, January 25; https://www.seradata.com/one-starlink-launch-batch-does-not-look-very-healthy-one-in-five-satellites-appear-to-have-failed/

[78] Davenport, C., *The Pentagon is looking for garbage collectors in space*, "The Washington Post", 2022, Janiary 30; https://www.washingtonpost.com/technology/2022/01/27/space-debris-pentagon-contract/

[79] Pultarova, T., *Commercial space clean-up service could be ready in 2024*, "Space.com", 2021, May 26; https://www.space.com/commercial-space-debris-removal-2024-astroscale

[80] Howlett, A., *OneWeb, Astroscale, and the UK and European Space Agencies Partner to Launch Space Junk Servicer ELSA-M with €14.8 million Investment*, "Astroscale", 2022, May 27; https://astroscale.com/oneweb-astroscale-and-the-uk-and-european-space-agencies-partner-to-launch-space-junk-servicer-elsa-m-with-e14-8-million-investment/

[81] *Internet penetration in Africa January 2022, by country*, "Statista", 2022; https://www.statista.com/statistics/1124283/internet-penetration-in-africa-by-country/

[82] *U.S. household dial-up internet connection usage 2019, by state*, "Statista", 2022; https://www.statista.com/statistics/185532/us-household-dial-up-internet-connection-usage-by-state-2009/

[83] Clery, D., *Starlink already threatens optical astronomy. Now, radio astronomers are worried*, "Science", 2020; https://doi.org/10.1126/science.abf1928

[84] NASA, *Shuttle Tire Sensors Warn Drivers of Flat Tires*, "NASA Spinoff", 2019; https://spinoff.nasa.gov/Spinoff2019/t_2.html

[85] NASA, *NASA Spinoff Brochure 2022* "NASA Spinoff", 2022, p. 64; https://spinoff.nasa.gov/

[86] NASA, *Fogless Ski Goggles*, "NASA Spinoff", 1976; https://spinoff.nasa.gov/node/9613

[87] United Nations, *The Sustainable Development Goals Report*, 2022; https://unstats.un.org/sdgs/report/2022/